OIL POLLUTION AS AN
INTERNATIONAL PROBLEM
A Study of Puget Sound and
The Strait of Georgia

OIL POLLUTION AS AN INTERNATIONAL PROBLEM
A Study of Puget Sound and The Strait of Georgia

WILLIAM M. ROSS

Western Geographical Series, Volume 6

Series editor Harold D. Foster

Department of Geography
University of Victoria
Victoria, British Columbia
Canada

1973 University of Victoria

Western Geographical Series, Volume 6

EDITORIAL ADDRESS

Harold D. Foster, Ph.D.
Department of Geography
University of Victoria
Victoria, British Columbia
Canada

ACKNOWLEDGMENTS

The author is grateful to the Canada Council for fellowships which enabled him to complete the manuscript while a doctoral student at the University of Washington. The book has been published with the help of a grant from the Social Science Research Council of Canada, using funds provided by the Canada Council.

Numerous individuals contributed in different ways to completion of the study. Without acknowledging their specific contributions, the author is deeply grateful to: Roger Aase, David Anderson, M.L.A., Ken Benjamin, Dr. William Beyers, Herb Buchanan, Dr. James A. Crutchfield, Paul Cunningham, Dr. Evan Denney, Dr. Harry Foster, Dr. Steve Golant, Tom Goode, M.P., Paul Kangas, Sid Martin, Dr. Richard Morrill, Dr. Phillip Muehrcke, Charles Ogrosky, Naomi Pascal, Lieut. James Phaups, Phil Rash, David Reid, Dr. Timothy O'Riordan, Tom Russell, Dr. Allan Sommarstrom, Randy Sorensen, Dr. Morgan Thomas, Dr. George Tomkins, Dr. Edward Ullman, Dr. Juris Vagners, Edward Weymouth, Ev Wingert, the Kearney and Taylor families and Martha Richards, Susan Barwell, and Jean Rystadt.

Whatever insight and clarity this study may now possess is due in large measure to the patience, encouragement, and intellectual challenge provided by Drs. Marion Marts and Richard A. Cooley. The author has benefited greatly from their scholarship and dedication to both issue and student.

Finally, family and friends gave much and generously endured a demanding author. For their many kindnesses, he remains grateful.

William M. Ross

Kent, Ohio
October, 1972

The Series editor would also like to thank the many people who have combined to facilitate the successful production of this volume. Miss Lorraine Atherton has spent many hours aiding with editing. Many of the photographs were either taken or solicited by Mr. Ian Norie who together with Mr. John Bryant produced the cover and book design. Miss Lorraine Image typed the final copy. The kind cooperation of all these individuals is very gratefully acknowledged.

Harold D. Foster

University of Victoria
Victoria, British Columbia
February, 1973

PREFACE

Studies purporting to evaluate public policy necessarily reflect the values, beliefs and background of the authors. Seldom, however, have these studies attempted to set out the views of the authors in a short statement preceding the evaluation. Most often the reader is expected to discern particular points of view from what has been written and often criticisms of the same evaluation are unfair because the reader injects his own value judgements into the criticism. While one may disagree with a particular view, the value of any study must be based on how well that point of view has been argued. What follows therefore is an initial attempt to outline this author's approach to environmental problems as the study is being conceived.

Governments in North America have traditionally devoted themselves to furthering economic growth. Some time in the 1960's many of the basic needs of the great majority of the populace were satisfied. Since then, these people, who are primarily urban and suburban citizens have had time and money to concern themselves with other interests ranging from recreation and the arts, to the deteriorating environment. Concurrently, many others - minorities, the aged, the poor - especially those who have failed to enjoy the benefits of the economy or who have felt that the biases in the system are neither adequate nor just to meet the real needs of the country and society have expressed their dissatisfaction.

These problems relate to economic growth and to institutions dealing primarily with economic questions in which compromise and tradeoffs are possible. Many of our current problems, however, concern issues which cannot be evaluated in sheer economic terms. Those who are dissatisfied with society question whether our institutions can cope with problems that are not inherently economic in nature and may in fact challenge the whole foundation of a society which has relentlessly pursued economic growth.

Environmental degradation is a major question facing society. It is a problem which must be solved over a long period of time. The emphasis should be shifted away from economic growth per se to other broad problems of society including the environment. This does not mean that we ignore the economy, but if economic interests conflict with other important priorities, then economic growth should be sacrificed.

Most of our approaches to environmental problems have been segmented and mission oriented, with little integration and long range planning. We have approached problems through institutions such as the United States Army Corps of Engineers, by constructing facilities to alleviate our problems. The Corps, for example, builds flood control structures ostensibly to reduce flood damage and yet in almost all flood plains damage potential has actually increased. A comprehensive approach to environmental problems is needed, an approach which can see how all things are integrated as well as the specialist approaches which we have relied on in the past.

Beyond these general beliefs about the place of environmental problems in society, specific biases concerning pollution may also appear. I believe the public should have, as a matter of course, the right to a pollution free environment. Those damaged by pollution, whether that damage be direct or indirect, should have speedy recourse to just compensation. Pollution expenses should, as a matter of principle, be borne by industry as a legitimate part of its production costs since it is better able to absorb and distribute costs.

Undoubtedly the experience of living in the region will influence this evaluation. No one who has travelled the coastal waters and enjoyed the magnificence of the islands and mountains can fail to appreciate the beauty of the Pacific Northwest. Any assessment of the threat to the environment of the region could undoubtedly be exaggerated by this observer who feels a deep attachment to the place.

The study may also have been clouded by the training and nation-
ality of the author. Geographers have traditionally viewed problems in
their areal context and have placed great emphasis on a regional approach
to human problems. The fact that much of this training was undertaken
in Canada may induce other values into the study. The national bias of
a Canadian, familiarity with a British parliamentary type of government
and a belief that a well-qualified civil service is capable of handling
most jobs after policy has been established may also have influenced the
conclusions.

Threats to the environment of the Pacific Northwest only exempli-
fy the damage that is being done to the global environment. Each incident
of pollution contributes to this decline. Ultimately some global forms of
management must be achieved if we are to arrest environmental deterioration; however, management at this level seems impractical at this time.
Improvements can and must be made but they will have to be more modest
in scope.

William M. Ross
October, 1972

TABLE OF CONTENTS

LIST OF TABLES

LIST OF FIGURES

LIST OF PLATES

CHAPTER 1
INTRODUCTION

> The challenge of the ocean is international as well
> as national in scope. As marine technology generates
> more activities and ambitions, nations must learn how
> to preserve marine resources as well as their respective
> tempers.... The ocean promises to be the ultimate chal-
> lenge to nations to co-exist in a watery planet whirling
> through space.[1]

Increased demand for oil in industrialized nations since World
War II has fostered an intensive exploration and drilling program in oil
producing countries. Most of the new wells have been land based, but
many new wells are being drilled offshore, thus increasing the potential
of major oil pollution in coastal areas. In addition, most of the industrialized nations of Western Europe and North America do not produce
enough crude oil to meet their own needs. Oil must be transported by
pipelines and tankers from the primary surplus regions in the Middle East
and South America as well as within countries. Oil spillage from offshore wells near California and Louisiana, from barges and small tankers
off the coasts of Newfoundland and Nova Scotia, from vessels in San
Francisco, and from numerous disasters involving large tankers off the
coasts of Europe and Africa have demonstrated that marine oil spillage
affecting coastal areas is rapidly becoming a serious international problem.

Many oil spills have taken place in international waters, and oil
from one spill can affect the coastlines of several nations. The ocean is
becoming important as an agent for absorbing and transferring pollution.[2]
It is possible to envisage a program to manage problems emanating from
marine oil spillage, but several factors hamper prospects of broad international management. First, the oceans are not alienated to one specific

1

owner. Second, nations or firms gain little economic benefit from controlling or preventing pollution when they are able to pass damages or treatment costs on to another nation. Third, pollution has only recently been
regarded as a global problem and a threat to the quality of the world environment. As a result little progress has been made in stemming oil pollution incidents which are occurring more frequently and occasionally
reaching catastrophic proportions. Before we can reverse this dangerous
trend we must first understand the dimension of our concern for the quality of the world environment at the national and international levels, and,
more specifically, the circumstances which give rise to marine oil pollution.

CONCERN FOR THE QUALITY OF THE WORLD ENVIRONMENT

For a major part of this century, pollution was regarded as a local or bilateral problem, but not as a threat to the quality of the world
environment. River pollution abatement has been the target of most international efforts, with air and ocean pollution receiving only scant attention until the 1960's. International commissions designed to protect
international waterways have formed the basis for existing cooperation to
control river pollution. The International Commission for the Protection
of the Rhine Against Pollution, the International Joint Commission (Canada and the United States), and the International Boundary and Water Commission (Mexico and the United States) are three typical examples. Each
has the power to investigate pollution problems and make recommendations,
but lacks authority to implement its findings.[3]

More than any other series of events, the fallout from the atmospheric testing of nuclear weapons and disastrous spills from large oil tankers demonstrated the global nature of the pollution problem. Fallout from
a nuclear device is distributed throughout the world in a matter of days.

2

Scientists argued that continued fallout could seriously endanger human health. As a result of much public outcry, the major nuclear nations agreed, in the partial Nuclear Test Ban Treaty of 1962, to desist from detonating nuclear devices in the atmosphere. Agreements to limit the effect of international oil pollution have been less comprehensive than those applying to atmospheric testing of nuclear weapons. One of the few detailed treaties is the International Convention for the Prevention of Pollution of the Sea by Oil of 1954, amended in 1962 (hereafter cited as the 1954 and 1962 Conventions). Under the 1954 regulations, convention ships were forbidden to discharge oil generally within 50 miles of a coastline and the 1962 amendments virtually prohibited discharge.[4] The Convention has no standards designed to reduce the number of tanker accidents and it is difficult to enforce in the absence of an intensive surveillance program. Nevertheless the treaties restricting nuclear testing and oil discharge do demonstrate that pollution is being increasingly recognized as a global problem. The United Nations attested to this concern by organizing an International Conference on the Human Environment in Stockholm in 1972.

While the international community has recognized pollution as an international issue, existing programs designed to combat pollution are not adequate to meet the threat. Ward and Boulding are concerned with the pace at which the international community recognizes global problems.[5] Barbara Ward argues that all "the irresistible forces of technological and scientific change are creating a single vulnerable, human community."[6] Until we harness these vast changes, meaningful solutions to global problems are not likely. Unless human solutions can be applied to problems created by science and technology, the very problems spell disaster for human society. Formation of a political community reflecting these changes is hampered by three great disproportions of influence; the first of power, the second of wealth, and the third of ideology.

Boulding has expressed similar concerns over the effect of science and technology but gives particular attention to their effects on the quality of the world environment. To Boulding, the present world economy exhibits characteristics of a "cowboy economy"; the cowboy economy being symbolic of boundless plains and resources and also "associated with recklessness, exploitative romantic and violent behavior which is characteristic of open societies."[7] Science and technology have been exploited to foster the present cowboy economy. If we are to arrest the decline in the quality of the world environment, future economic policies will have to be altered to reflect a "spaceman economy." In this future utopia the earth will become a[8]:

> single spaceship without unlimited resources of anything, either for extraction or pollution and, in which therefore, man must find his place in a cyclical ecological system which is capable of continuous reproduction of material form even though it cannot escape having inputs of energy.

Both Ward and Boulding view unharnessed science and technology as imminent dangers that require political solution on a global scale.

Hardin, while lamenting some of the disasters which have occurred as a result of offshore oil spills, contemplates a fast changing world moving to meet the problems resulting from science and technology. Paramount among the changes is growing rejection of the notion that an individual has a right to pollute excessively[9]:

> The heresy that no one has a right to pollute the media of the world (water, air and ether; the medium of light and radio waves) is changing to orthodoxy. Sooner or later the ecological ethic will prevail. Sooner or later industries will be forced to internalize so-called externalities.

Marx is also optimistic about the ability of man to correct the abuses he has perpetrated upon the oceans, but argues that past abuses must be rapidly corrected to ensure permanence of the oceans. He possesses no

delusions about the challenge. Obstacles such as a desire for exploitation, abuses by unproven technology and human carelessness will hamper
efforts to improve the quality of the ocean environment. The real challenge however is political; a challenge that ultimately depends on the
willingness of nation states to limit national desires and surrender sovereignty to benefit the global community.

The national political responses to international pollution have
been critically examined by Livingston, Wolman, and Ross.[11] All agree
that states have failed to anticipate international ecological needs and
respond positively to them only after conflict or crisis. This attitude reflects not only public apathy and lack of funds, but, most importantly, a
reluctance to surrender political sovereignty to an international organization. No organized constituency exists at the international level to press
for anti-pollution laws. Additional delays are inherent in getting a large
number of states to agree on complex legislation on a subject whose full
ramifications may not be completely understood. Most anti-pollution a-
greements that have been ratified are bilateral agreements designed to alleviate specific problems. When multilateral treaties have been consummated, they commonly lack sufficient powers to enforce an efficient management program. While all three authors recognize the limits of present
international agreements, they view, with varying degrees of optimism,
the evolution of an international ecological conscience as a postive first
step towards improving the quality of the world environment. Broadly
based anti-pollution regulations would form one essential control in improving environmental quality.

If the optimism expressed above is to prove legitimate, more intensive efforts will be needed to mesh scientific and technical expertise with
a political organization which has power to legislate and enforce needed
changes. Fay argues that an understanding of respective contributions
from science, technology and public management is a necessary prelude

to an integrated plan to secure abatement of oil pollution.[12] Crowe envisages two specific roles for science within such an integrated political organization. First, science could concentrate on developing technological innovations which would alleviate the problem. Second, science could utilize environmental monitoring devices to detect continued oil spills. This would ensure that the political organization had the necessary data for enforcing regulations and would place additional responsibility upon scientists to appraise the appropriateness of their own technology.[13] Social scientists would face considerable responsibilities in defining the ecological, social and economic needs within the political organization. Such definitions would entail an assessment of the respective national needs, especially when those needs conflict with international interests.

The ultimate challenge is faced by the nation state which must evaluate how it can best contribute to improving the quality of the world environment. Within most states, concern has been expressed over the quality of the global environment, but this concern is not always shared by national and sub-national governments. Some nations feel that well designed national programs with limited international obligations can lead to more efficient and equitable international pollution control. Other states have seen the world wide pervasiveness of problems such as oil spillage, and realize the need for an international organization to assess competing needs and recommend remedial actions. In either case the ultimate success of pollution abatement programs will depend largely on the willingness of states to surrender sovereignty to organizations of the type described above, organizations which have power to design, implement and enforce anti-pollution programs.

THE PROBLEM OF OIL POLLUTION

The potential sources of oil pollution are myriad, but some are more potentially threatening to the oceans than others. Natural seepage

from the earth and from sunken vessels and careless and inadequate disposal methods for used motor oil and industrial oil are common but do not constitute a major source of international ocean pollution. Most of these problems are either beyond man's management ability or of such a localized
nature that they are most effectively solved by municipal and other subnational authorities. Pipelines and storage facilities have been the source
of major oil spills, but the effect of these incidents has been largely limited to pollution of the land. Lines that have ruptured and polluted sur-
rounding waters have usually been associated with pipes connecting offshore oil rigs to onshore storage installations. The greatest threat to the
oceans stems from two sources. One is the move to offshore drilling for
oil. Offshore wells have been struck by blowouts; explosions have occurred on oil platforms; inadequate safety devices have been installed; and
ships have hit oil rigs. The second, and perhaps more important, is the
international trade in oil. Tankers of increasing size have frequently
been involved in collisions, groundings, and founderings which have polluted international waters and affected the coastlines of adjacent countries.

A complete analysis of all sources of international pollution is beyond the scope of this study. Instead the study will concentrate on international pollution caused by oil spillage from offshore oil rigs or from
large tanker operations and mishaps. Such a concentration will limit the
study largely to spills of considerable magnitude. Pollution emanating
from such sources as bilge pumping, while considerable, will not be considered. By limiting the sources considered it is thus possible to focus on
the damage caused by products usually produced by wells and carried by
tankers and other large commercial vessels. The study will therefore concentrate on pollution from persistent oil[14] (such as crude oil, diesel fuel,
and heating oil), and refined products (such as gasoline and kerosene),
and their ability to cause considerable environmental, commercial, aesthetic, and recreational damage.

Pertinent data on the damages resulting from large oil spills is largely descriptive, but it does provide an indication of the range of consequences that can result from a spill.[15] When spilled, oil is a potential floating drifting nuisance, capable of covering large expanses of water in a relatively short time. Each spill is unique, but there is no doubt that oil inflicts damage. Some damage, especially surface damage, is quite obvious and costs can be placed on the damage; however, much damage is hidden and long term (especially that to the subsurface marine and coastal ecosystems) and damage costs are therefore not fully known.

Spilled oil is often a graveyard for species of animal and plant life.[16] Birds, especially those frequenting coastal areas, are particularly vulnerable. They land on the oil mass, the feathers become matted and oil soaked and the birds often drown through lack of buoyancy. If they do not drown, toxicosis from ingested oil and exposure through loss of body heat can also result in death. Gulls and other seabirds are affected by oil, but they suffer much less damage than auks and a number of sea ducks. Even though the gulls are not as vulnerable, a spill from the fuel tanks of the frigate Seagate wrecked in 1956 off the Olympic Peninsula, Washington resulted in a mortality rate for gulls of 56.5 per mile of shoreline.[17] Fish are also affected. If the oil coated gills do not cause death, the flesh of the fish often absorbs the taste and odor from the oil and renders the species unfit for human consumption. Similarly, marine plant life such as seaweed also acquires a disagreeable taste and cannot be used.

Oil has an adverse effect on the food chain, but the full range and extent of damages are not fully understood or known. Some of the hydrocarbons in oil can be degraded by marine bacteria, but the process requires oxygen which is necessary for other forms of marine life. Other components of oil such as the aromatic content are potentially more dangerous. These components are incorporated into the food chain at a very low level and can become toxic in greater concentrations at higher levels to the extent of contaminating marine food supplies. Dr. Blumer of Woods Hole has even linked such components to sources of cancer when incorporated in the food chain.[18] In addition marine life manufacture natural hydrocarbons which are essential to marine organisms. Salmon that move inland to spawn are thought to be guided to their spawning grounds by small traces of natural hydrocarbons. An imbalance in the amount of hydrocarbon potentially threatens the ability of salmon to successfully complete their life cycle and safely return to their spawning rivers and streams.[19]

While spilled oil inflicts damage on the marine environment, efforts to clean up the spill often cause much more destruction. Experiences in South Africa have demonstrated that spraying chemical detergents from ships is the most effective way to treat oil slicks, although the detergents were highly toxic to marine life. In the Torrey Canyon case, studies showed oil itself accounted for destruction of 30 percent of the plankton while detergents used in attempting to dissipate the oil killed off 95 percent of the remaining plankton along the British coast.[20] All detergents produced harmful effects on fishes, crustacea and mollusca, the toxicity increasing with the effectiveness of the detergent.[21]

Depending on the season and location, the amount of oil spilled can vitally affect man and his economy. Fishermen often cannot fish during oil spills and if they do their catch is sometimes unfit for consumption and their gear is fouled beyond repair. Hotel and restaurant businesses catering to fishermen or those in tourist areas incur noticeable losses of revenue due to contamination of the open waters and beach areas. Those patrons who remain to observe spills often cause restaurants and hotels to bear the burden of cleaning oil soiled rugs and furniture. If the spill occurs in a bay, port, or estuary, many small pleasure boats can be coated with oil.

Besides the costs and losses that are suffered by private individuals and businesses, governments incur heavy expenses in cleaning up fouled

beaches, wharves, and adjacent waters. In the Torrey Canyon accident,
Great Britain and France claim to have spent $12 million in cleanup oper-
ations alone. Preliminary estimates indicate that the Canadian government will spend over $4 million to recover and clean up oil from the Arrow.

In addition to direct monetary expenses, man also experiences inconvenience and annoyance. Odor from a spill often plagues nearby residents. Water sports such as swimming and boating must be curtailed.
Beaches are fouled and coastlines become black and ugly. Aesthetic costs
are hard to measure, but they are no less real to individuals who value the
beauty of the coast and sea.[22]

OIL POLLUTION AND THE WORLD ENVIRONMENT

Ever since oil became so necessary to industrial societies, isolated
incidents of oil spillage have caused localized problems. As oil drilling
and transportation increased, so did the frequency of mishaps and the a-
mount of oil spillage. The media focused attention on the world wide consequences of oil spillage and public concern for a "quality life" pressed
governments to consider remedial actions. While oil pollution has been
present for a long time, the international dimensions of the oil spillage
problem are largely a post-World War II phenomenon and, like the growth
in drilling and transportation, are not likely to diminish in the immediate
future.

In the last twenty years the world has experienced a tremendous
surge in the demand for energy. The increase in populations, the rising
standard of living, and the growth of industry are all contributing to the
swelling demand for energy. Various studies, while differing in their
projections, all conclude that for the present and foreseeable future there
is no adequate substitute for crude oil. Hartshorn (see Table 1) outlines
the rapid shift in world energy since 1930 away from solid fuels such as
coal to oil and natural gas. Oil and gas now account for over 55 percent

PLATE 1a
The sinking of the Arrow

PLATE 1b
Vanlene aground on The
West Coast of Vancouver
Island

Photo: A.B. Ages

11

of world energy consumption of commercial fuels. The demands in the free world for oil and gas are even greater. The Report of the Panel on Marine Resources suggests that well over 60 percent of the free world's demand for energy will be met by oil and gas (see Table 2). While the energy mix has shifted considerably, the absolute demands for energy are even more striking. Production of petroleum and natural gas will have to double between 1965 and 1985 if we are to meet projected demands.

Projection of petroleum consumption and demand for Canada and the United States parallel the pattern for the free world. Landsberg,

TABLE 1

WORLD ENERGY CONSUMPTION, COMMERCIAL

FUELS, ORDERS OF MAGNITUDE

Fuel Type	1929	1937	1950	1965	1971
	Million Metric Tons Coal Equivalent (percentages in brackets)				
Solid Fuels	1367(80)	1361(75)	1569(62)	2260(42)	2698(38)
Hydro-Electricity	14(1)	22(1)	41(2)	114(2)	166(3)
Natural Gas	76(4)	115(6)	273(11)	892(17)	1309(18)
Oil	255(15)	328(18)	636(25)	2118(39)	2951(41)
Nuclear Electricity	–	–	–	3	23
Total	1712(100)	1826(100)	2519(100)	5387(100)	7147(100)

Source: J. E. Hartshorn, Politics and World Oil Economics. New York: Frederic A. Praeger (1967), p. 389.

TABLE 2

FREE WORLD ENERGY DEMANDS BY COMPONENT FUELS

Fuel Type	1950	1965 estimated	1975 estimated	1985 estimated
	Millions of barrels per day of oil energy equivalent (percentages in brackets)			
Petroleum	10.1(32.8)	25.9(46.7)	40.1(49.5)	56.6(48.8)
Natural Gas	3.0(9.9)	8.6(15.5)	14.1(17.4)	22.2(19.1)
Coal	14.8(48.3)	15.6(28.1)	18.4(22.7)	21.4(18.4)
Water Power	2.1(6.7)	4.1(7.4)	5.8(7.2)	7.6(6.6)
Nuclear	-	.2(.4)	1.8(2.2)	7.6(6.6)
Other	.7(2.3)	1.1(1.9)	.8(1.0)	.6(.5)
Total	30.7(100.0)	55.5(100.0)	81.0(100.0)	116.0(100.0)

Source: Commission on Marine Science, Engineering and Resources. Marine Resources and Legal-Political Arrangements for Their Development, Vol. 3, Report of the Panel on Marine Resources Panel VII. Washington: United States Government Printing Office (1969), p. Vii-193.

Fischman, and Fisher conducted the most exhaustive examination of United States requirements to the year 2000. Their study showed (see Table 3) at least a 100 percent increase in the demand for petroleum between 1960 and 2000 depending on the projection figures. The highest projection anticipated over a 500 percent increase in demand. The absolute demand for petroleum will increase substantially, while at the same time oil and natural gas (see Table 4) will continue to account for between 60 percent and 70 percent of the total energy consumed in the

TABLE 3

PROJECTION OF TOTAL PETROLEUM DEMAND IN ALL
USES TO THE YEAR 2000 FOR THE UNITED STATES

Projection	1960	1970	1980	1990	2000
			(in billion barrels)		
Actual	3.19	–	–	–	–
Low Projection	–	3.61	4.44	5.56	7.11
Medium Projection	–	4.08	5.34	7.17	10.03
High Projection	–	4.63	6.72	10.16	16.00

Source: Hans H. Landsberg, Leonard L. Fischman, and Joseph L.
Fisher. Resources in America's Future. Baltimore: Johns
Hopkins Press (1963), p. 849.

TABLE 4

UNITED STATES HISTORICAL AND PROJECTED
CONSUMPTION OF ENERGY IN PERCENTAGES OF TOTAL
(MEDIUM PROJECTION)

Energy Type	1940	1950	1960	1970	1980	1990	2000
				percent			
Coal	50.4	37.7	24.5	21.6	19.9	16.2	13.3
Oil	31.8	36.6	39.1	37.6	37.2	38.2	40.5
Natural Gas	13.3	19.9	29.2	31.8	30.5	27.8	25.0
Natural Gas Liquids	.9	2.0	3.6	4.2	4.3	4.6	5.1
Hydro Power	3.6	4.4	3.6	4.2	3.4	2.7	2.1
Nuclear Power	–	–	–	.6	4.7	10.1	14.0
Total	100.0	100.0	100.0	100.0	100.0	100.0	100.0

Source: Hans H. Landsberg, Leonard L. Fischman and Joseph L.
Fisher, Resources in America's Future. Baltimore: Johns
Hopkins Press (1963), p. 858.

TABLE 5

ENERGY CONSUMPTION IN CANADA BY TYPE,
1926, 1953, AND 1980, ESTIMATED

Energy Type in terms of tons of coal equivalent	percentage of energy consumed		
	1926	1953	1980 estimated
Coal	69.0	39.0	16.0
Petroleum	10.0	42.0	45.0
Natural Gas [a]	2.0	4.0	25.0
Wood	16.0	7.0	1.0
Water Power [b]	3.0	8.0	11.0
Nuclear Energy [b]	-	-	2.0
Total	100.0	100.0	100.0

a Including natural gas liquids
b Measured in terms of its contributions as electricity

Source: John Davis, Canadian Energy Prospects. Ottawa:
Queen's Printer, (1957), p. 3.

United States. By 1980 Davis estimates (see Table 5) that petroleum and natural gas will account for 70 percent of Canada's energy needs. These figures are comparable to United States projections for the same period. A later study by the National Energy Board, however, estimates that petroleum and natural gas will account for between 55 percent and 65 percent of the total Canadian energy requirements between 1966 and 1990 (see Table 6). More of the Canadian energy demand will be met by hydroelectricity and nuclear power than is the case in the United States. In an absolute sense petroleum production is expected to increase from

TABLE 6

RELATIVE IMPORTANCE OF CANADIAN PRIMARY
ENERGY REQUIREMENTS

Energy Type	1966	1975	1980	1985	1990
			percent		
Petroleum Fuels	45.9	41.2	39.7	37.7	35.9
Natural Gas	16.3	19.6	20.1	20.4	20.0
Coal and Coke	12.5	11.6	13.1	11.9	10.9
Hydro Electricity	25.3	25.2	22.8	21.9	18.9
Nuclear	-	2.4	4.3	8.1	14.3
Total	100.0	100.0	100.0	100.0	100.0

Source: National Energy Board, Energy Supply and Demand in Canada
and Export Demand for Canadian Energy, 1966 to 1990.
Ottawa: Queen's Printer (1969), p. 175.

10 percent to 15 percent per year on a 1969 base of 1.3 million barrels
per annum.[23]

   To meet the demands for, and consumption of, oil there have been
intensive programs of exploration and drilling throughout the world. These
programs have resulted in new discoveries of oil in environments that make
drilling difficult and in places far removed from the markets. Between
1954 and 1969, for instance, 8,000 offshore wells were drilled in the
United States; and if the present rate continues, 3,000 to 5,000 wells will
be drilled annually by 1980.[24] The emergence of offshore drilling and
increased production in the Middle East and Venezuela coupled with increased demand in Europe and North America has necessitated transportation of oil by tanker as well as by pipeline over greater and greater

distances in ever larger vessels.

To transport oil the industry is presently using many vessels of
World War II vintage with carrying capacities of 20,000 deadweight tons
(DWT). There are many tankers, however, with capacities of 100,000
DWT, vessels of 200,000 and 300,000 DWT are being constructed, and
tankers of 450,000-500,000 DWT are thought to be possible in the near
future.[25]

The term "T-2" is a common means of measuring tanker capacity.
The T-2 tankers were constructed by the United States Maritime Commission during World War II and had a capacity of 17,765 DWT. Between
1958 and 1968 the number of tankers increased from 3,146 to 3,748; however, in T-2 equivalents world tanker capacity has grown from 3,403 to
8,202 in the same period.[26] This increase in the size of tankers poses
immediate problems. Many harbors are unable to handle large tankers.
The Manhattan, for instance, has a draft of 50 feet, but there is no place
in the United States where she can work at full draft except Long Beach,
California and Puget Sound. Elsewhere the ship must be topped off or
lightened at sea. As a result, superports have been or are being constructed,
and plans have been made to dredge and maintain some channels for
300,000 ton ships laden to 75 feet draft and to develop less costly free-
swinging open sea tanker moorings.[27] The move to large tankers increases
the hazard from any single incident and the possibility grows that a single
spill from one of these large tankers might impair the maritime resources
of several nations.[28]

The problems of oil spillage were probably best demonstrated by
the grounding of the Torrey Canyon, but this incident is only representative of numerous spills from both drilling and tanker foundering. A study
by Battelle Memorial Institute succinctly summarized the Torrey Canyon
disaster:[29]

PLATE 2
The <u>Manhatten</u> in Arctic waters

Photo: Canadian Petroleum Asso

> The Torrey Canyon (118,000 deadweight tons) disaster
> off Lands End, England starting on March 18, 1967
> and continuing for well over a month, demonstrated
> with dramatic clarity man's ability to pollute large
> expanses of the ocean and adjacent shores to an extent
> having international as well as large financial implica-
> tions. This disaster ... pointed up the lack of a well
> assembled body of knowledge and experience that could
> be applied. As a result defensive measures were insti-
> tuted largely on an "ad hoc" basis and in an atmosphere
> of emergency.

These conclusions could well be made for the most serious spills that have
occurred in North America. The Torrey Canyon disaster was viewed with
distant concern in North America, but the large spill from a blownout
well off the Santa Barbara coast, the threat posed to the Arctic by the voyage of the Manhattan, and the sinking of the Arrow off the Nova Scotia
coast demonstrated that oil spillage was a large and immediate problem
for both the United States and Canada.[30]

The most recent published data on oil spillage for the United
States is for 1965 and 1966. The United States Army Corps of Engineers
estimates that there were over 2,000 spills in the United States in 1966 of
which 60 percent came from marine based facilities and 40 from land based
operations.[31] The maritime casualty rate is reflected in Table 7 which
summarizes the record of United States registered vessels world-wide and
foreign flags in United States waters. These figures fluctuate, but are at
a level sufficiently high to justify serious concern. Casualty figures for
Canadian ships and foreign vessels in Canadian waters are not published,
but would be proportionally parallel for accidents in Canadian waters and
less for Canadian registered vessels engaged in world-wide trade.

The serious concern expressed over the number of oil spillage incidents in North America reflects a metamorphosis in our attitude towards
the environment. Isolated incidents were once regarded as an unfortunate
but perhaps necessary consequence of economic growth. Great faith was

PLATE 3a
Oil from the Arrow fouling shore
line at Arichat Harbour

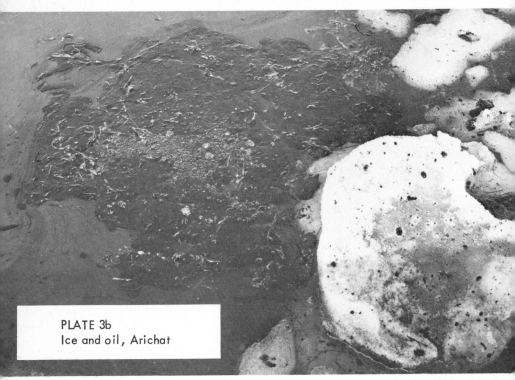

PLATE 3b
Ice and oil, Arichat

Photo: Fed. Dept. of Transp

20

PLATE 4a
Brush boom at West Arichat containing Arrow oil

PLATE 4b
Oilevator in operation cleaning the shore at River Inhabitants

Photo: Fed. Dept. of Transport

21

TABLE 7

UNITED STATES MARITIME CASUALTY RATES

	Fiscal Year 1966	Fiscal Year 1967
Number of Casualties – All Types	2,408	2,353
Vessels over 1,000 tons	1,310	1,347
Tank Ships and Barges	470	499
Locations: U.S. Waters	1,685	1,569
Elsewhere	723	784
Types of Casualties:		
Collisions	922	1,090
Explosions	175	168
Groundings with Damage	302	282
Founderings, Capsizings, Floodings	315	230

Source: United States Departments of Interior and Transportation.
Oil Pollution: A Report to the President. Washington, D.C.:
United States Government Printing Office (1968), p.8.

placed in the ability of technology to solve immediate problems created by
oil spills and to arrest the deteriorating quality of the environment. In the
1960's much of this conventional wisdom about man's relationship towards
the environment began to be challenged by the popular media, politicians,
and academics.[32] The sources of this attitudinal change are complex, but
concern for a deteriorating environment has been linked, though not limited to, societies which have satisfied their basic needs for survival, experienced a shortage of open space, and attained a considerable degree of
technological advancement and urbanization.[33] Whatever the source of
this change, governments have been pressed into action, albeit on a limited scale. In the United States the Water Quality Improvement Act of

1970 was passed to help prevent and control oil spills. States such as Maine and Washington have enacted laws designed to protect their coasts from oil spills.[34] In Canada a unique pollution control zone has been established for the Arctic, and amendments to the Canada Shipping Act have been passed to tighten Canadian control over the operation of vessels in Canadian waters (see Chapter Three). While attitudes toward oil spillage have changed and some governments have acted, there remains the question of whether these developments are adequate to cope with the international dimensions of the problem.

OIL POLLUTION AS AN INTERNATIONAL PROBLEM IN THE PUGET SOUND AND STRAIT OF GEORGIA REGIONS

Actions by national governments and multilateral organizations have been singularly ineffective in preventing and controlling global oil pollution. Domestic laws of one nation can at best control actions of its own nationals world-wide, and establish oil spillage liabilities for all within that nation's territorial waters. When oil spills occur on the high seas or in the waters of one state and affect another, liability must be established through some form of international agreement or organization. Efforts have been made to control international oil pollution through the 1954 and 1962 Conventions; however, the number of oil spillage incidents from tankers and offshore oil rigs continue to increase. Conferences in London (1967) and Brussels (1969) under the auspices of the Intergovernmental Maritime Consultative Organization (IMCO) have focused attention on preventive measures such as designation of sea lanes and prohibited areas, shore guidance systems for ships near land, and organization of international watches, but there is no legislative body or regulatory agency to adopt and enforce standards. While significant advances were made towards new international conventions at these conferences, no comprehensive agreements emerged.[35] There is no organized constituency at

the international level pressing to improve the quality of the world environment by preventing and controlling oil pollution. National governments must represent ship owners, oil producers, cargo owners, oil con-
suming interests, terminal owners, ship personnel, and ship yard owners
as well as pursue environmental quality. International agreements to control oil pollution reflect these conflicting and diverse interests. As a result, few stringent controls have been imposed on industry without their
various consents, and international organizations such as IMCO remain
consultative bodies with limited powers.

     This reluctance to entrust international bodies with greater powers
has caused some nations who have been threatened by oil pollution to consider stringent national controls. A report to the Canadian Ministry of
Transport recommends:[36]

> ... that consistent with initiatives taken by the
> Government [Canadian] with respect to Arctic
> Pollution and at the IMCO special conference on
> pollution in 1969, Canada take a parallel initiative
> to convene a conference of all those concerned to
> write a new international convention for the organ-
> ization and control of shipping throughout the world...
> We do not underestimate the patience and persever-
> ance that will be necessary to achieve the foregoing
> international accord, but the substantial elimination
> of oil pollution we suggest is an urgent national
> matter. Canada should enact laws and establish
> practices to protect Canadians and lay down standards
> for the transportation and handling of petroleum
> products which will effectively bring some much
> needed discipline into this area.

In the immediate future we shall have to rely largely on efforts at the national and perhaps the bilateral level to curb immediate problems of international pollution. An examination of Canadian and American attempts
to control international pollution close to their own borders will serve as
a good test of international desires to prevent and control oil pollution.

One, Canada, is a nation with a long coastline and a small merchant
fleet; the other, the United States, has a similarly extensive coastline,
but also has considerable merchant fleet and defense interests which are
concerned with retaining current concepts of freedom of the seas. Relations between the two countries have been cordial and mechanisms do
exist (in bodies such as the International Joint Commission) for dealing
with pollution problems.[37] If these two nations are unable to deal with
pollution problems, then prospects for broader international controls will
be bleak.

Caldwell has stated that, until recently, the pressure of man on
his environment has been much less in some countries than in others. This
situation is now changing and comparative studies of nation-state experiences with natural resource issues will become increasingly important for
all countries.[38] Caldwell also might have alluded to the value of such
studies as they relate to comparative problems of resource management
when the problem is international in scope. International problems over
air and water pollution must be considered at both the national and international levels if we are to arrest the decline in the quality of the world
environment, although the most productive efforts are likely to be at the
national level in the immediate future. Much can be learned from one
country's efforts to control pollution, but there are distinct dangers in
world-wide adoption of such programs because each society has differing
social, economic, political, and cultural characteristics and goals.

Examining a particular problem of pollution between Canada and
the United States may provide insights into how individual nations might
act unilaterally to reduce the amount of international pollution. It may
also provide insights into the types of unilateral organizations which can
prevent and correct pollution problems, and the amount of responsibility
assigned to a regulatory international agency and to the nation-state.

Care must be taken in comparing resource problems and solutions in Canada and the United States because the two federal systems are based on differing premises.[39] If the value of comparative analysis of resource problems is to be realized, we must understand how to achieve stated management goals within each political system and within the international community. This understanding is based on the manner in which each federal system deals with environmental problems.

The Pacific Northwest is a good study area because the potential for oil spillage in the region is increasing and thus provides a good test of international policy.[40] Domestic oil trade has increased significantly. The discovery of oil on the north slope of Alaska along with the development of superports such as that at Roberts Bank, south of Vancouver, B.C., could mean an increase in the number and size of tankers and dry cargo vessels frequenting coastal waters in Washington and British Columbia. In addition, permits have been issued for oil exploration off all of the Pacific coast and drilling has commenced in some areas. Given that oil spillage is an increasing international problem and that the potential for this hazard will be increasing in the Puget Sound and Strait of Georgia regions,[41] it is proposed to examine the broader question of adjustment in governmental jurisdictions within the context of this threat.

The major purpose of this study is to examine methods by which management of international resource problems can be improved. More specifically it will examine in a specific case, the federal systems of Canada and the United States as they apply to particular problems, and suggest adjustments which may be necessary to realize closer relationships between the jurisdictions of existing governmental units at the federal, state, and provincial levels and the areal scale of environmental problems.

The study will also critically analyze legislation and regulations in Washington and British Columbia which deal with the oil spillage problem. This analysis will focus on four questions: (1) What insights can the literature on externalities, common property resources and water quality control provide for the improvement of management programs to prevent and control oil pollution? (2) Is the current legislation adequate to cope with the oil spillage problem now and in the future? (3) How closely related are state and provincial regulations, and what is the nature of the integration of these regulations with respective federal legislation? (4) Are there any provisions in existing regulations which would facilitate international management? Utilizing results of this critical analysis, an attempt will be made to evaluate the strengths and weaknesses of each federal system with regard to an important problem of environmental quality.

The study will attempt to suggest programs and administrative changes which would encourage better management of problems such as oil spillage. In the chapter that follows, the problem of oil spillage near an international boundary is examined within the conceptual framework provided by externality, welfare, and common property resource theory, and suggestions are advanced for evaluating existing programs to prevent and control oil pollution. In subsequent chapters specific changes are suggested for the study area and the applicability of these administrative and regulatory changes to other regions of Canada and the United States are evaluated. Attempts to prevent oil spillage in other parts of the world are investigated. A complete and systematic analysis of all these efforts is beyond the scope of this study. Nevertheless, based on conclusions reached through the study, changes in international organizations are suggested which might improve their management programs.

REFERENCES

1. MARX, W. The Frail Ocean. New York: Ballantine Books (1967), p. 7.

2. FAY, A. J. "Oil Spills: The Need for Law and Science, "Technology Review, No. 72 (January 1970), p. 33.

3. WOLMAN, A. "Pollution as an International Issue, "Foreign Affairs, No. 47 (October 1968), pp. 167-168.

4. LIVINGSTON, D. "Pollution Control: An International Perspective," Scientist and Citizen, No. 10 (September 1968), pp. 172-182.

5. WARD, B. Spaceship Earth. New York: Columbia University Press (1966); see also BOULDING, K. E. "The Economics of the Coming Spaceship Earth" in JARRETT, H. (ed.) Environmental Quality in a Growing Economy. Baltimore: Johns Hopkins Press (1966), pp. 3-14.

6. WARD, op. cit., p. 1.

7. BOULDING, op. cit., p. 9.

8. Ibid.

9. HARDIN, G. "Finding Lemonade in Santa Barbara's Oil," Saturday Review, No. 52 (May 10, 1969), p. 21.

10. MARX, op. cit., p. 253.

11. LIVINGSTON, op. cit.; WOLMAN, op. cit., pp. 164-175; ROSS, W. "The Management of International Common Property Resources," Geographical Review, No. 61 (July 1971), pp. 325-338.

12. FAY, op. cit., p. 33.

13. CROWE, B. "The Tragedy of the Commons Revisited" Science, No. 166 (November 28, 1969), p. 1107.

14. Persistent oils are characterized by their inability to dilute rapidly or readily in water and their relative stability and buoyancy compared to refined products.

15. For a good description of the consequences of one spill, the oil discharged from the wreck of the Liberian tanker Arrow on Cerberus Rock in Chedabucto Bay, Nova Scotia, on February 4, 1970, see Canada, Parliament, House of Commons, Special Committee on Environmental Pollution. Minutes of Proceedings and Evidence , Issues Numbers 3 and 4, November 5 and 10, 1970. Ottawa: Queen's Printer (1970).

16. BELLAMY, D.J. "Effects of Pollution from the Torrey Canyon on Littoral and Sub-littoral Ecosystems," Nature, No. 216 (December 23, 1967), pp. 1170-1173; HAWKES, A. L. "A Review of the Nature and Extent of Damages Caused by Oil Pollution at Sea," Transactions, North American Wildlife Conference, No. 26 (1961), pp. 343-355; NELSON-SMITH, A. "The Effects of Oil Pollution and Emulsifier Cleansing on Shore Life in Southwest Britain," Journal of Applied Ecology , No. 5 (April 1968), pp. 97-107; SCHACHTER, O. and SERWER, D. "Marine Pollution Problems and Remedies," American Journal of International Law, No. 65 (January 1971), pp. 88-95.

17. International Conference on Oil Pollution of the Sea, October 7-9, 1968 at Rome. Proceedings. Winchester: Wykeham Press (1968), pp. 82-83.

18. THORNE, C.M. "How Can the People of the State of Washington Coexist with the Oil Industry?," Report submitted to the Oceanographic Institute of Washington (December 1, 1970), pp. 6-11.

19. FAY, op. cit., pp. 33-34.

20. YOSHIOKA, T. "Problems of International Control of Oil Pollution of the Sea," unpublished paper, School of Law, University of Washington (undated), p. 3.

21. BOYLE, C. L. "Oil Pollution of the Sea: Is the End in Sight?," Biological Conservation, No. 1 (July 1969), pp. 321-322.

22.   United States, Departments of Interior and Transportation.
       Oil Pollution: A Report to the President. Washington:
       United States Government Printing Office (1968),
       p. 4.

23.   INGLIS, L. "Growth Rates for Oil and Natural Gas will Run
       10%-15% for Years," Financial Post, No. 64 (October 3,
       1970), pp. 13-18.

24.   United States, Executive Office of the President. Offshore
       Mineral Resources; A Challenge and an Opportunity,
       Second Report of the President's Panel on Oil Spills.
       Washington: United States Government Printing Office
       (1969), p. 3.

25.   SWEENEY, J.C. "Oil Pollution of the Oceans," Fordham Law
       Review, No. 37 (1968-1969), p. 157.

26.   TUCKER, A. J. "Boom in Tankers Ahead," Ocean Industry,
       No. 5 (January 1970), p. 35; see also COOKE, R. F.
       "Oil Transportation by Sea" in HOULT, D.P. (ed.)
       Oil on the Sea. New York: Plenum Press (1969),
       pp. 103-112.

27.   TUCKER, op. cit., p. 36; LITTLE, C. H. "Giant Bulk Carriers,"
       Canadian Geographical Journal, No. 77 (December 1968),
       pp. 196-203.

28.   SWEENEY, op. cit., p. 157; United States Departments of
       Interior and Transportation, op. cit., p. 6;
       RANKEN, M. B. F. "Can We Delay the Next
       Major Tanker Disaster?," Ocean Industry, No. 6
       (June 1971), pp. 35-39.

29.   Battelle Memorial Institute, Pacific Northwest Laboratories. Oil
       Spillage Study: Literature Search and Critical Evalua-
       tion for Selection of Promising Techniques to Control
       and Prevent Damage; to Department of Transportation,
       United States Coast Guard, Washington, D.C.
       Washington: United States Department of Commerce,
       Clearinghouse for Federal Scientific and Technical
       Information (1967), p. 1-1; see also COWAN, E.
       Oil and Water: The Torrey Canyon Disaster. Phila-
       delphia: Lippincott (1968); and BONE, Q. and
       HOLME, N. "Lessons from the Torrey Canyon ,"

New Scientist, No. 39 (September 5, 1968), pp. 492-493.

30. HOLMES, R. W. "The Santa Barbara Oil Spill," in HOULT, D.P. (ed.) Oil on the Sea. New York: Plenum Press (1969), pp. 15-28; HARDIN, op. cit., pp. 18-21; GALWAY, M. "What We Can Win in the Arctic," Saturday Night, No. 85 (April 1970), pp. 23-25; McTAGGART-COWAN, P. D., SHEFFER, H. and MARTIN, M. A. Report of the Task Force-Operation Oil. 3 vols. Ottawa: Ministry of Transport, Information Canada (1970).

31. United States, Departments of Interior and Transportation, op. cit., p. 6.

32. COOLEY, R.A. "Introduction: Politics, Technology and the Environment," in COOLEY, R. A. and WADDESFORDE-SMITH, G. (eds.) Congress and the Environment. Seattle: University of Washington Press (1970), pp. ix-xix.

33. GALBRAITH, J. K. The Affluent Society. Toronto: Mentor Books (1963); WHYTE, W. The Last Landscape. New York: Doubleday (1968).

34. Conservation Foundation. Conservation Newsletter (November 1970), pp. 10-11.

35. O'CONNELL, D. M. "Reflections on Brussels: IMCO and the 1969 Pollution Conventions," Cornell Journal of International Law, No. 3 (Spring 1970), pp. 161-188.

36. McTAGGART-COWAN, SHEFFER, and MARTIN, op. cit., Vol. 1, pp. 28-29; for a more detailed discussion of Canadian efforts to control pollution, especially the threat in the Arctic, see KONAN, R. W. "The Manhattan's Arctic Conquest and Canada's Response in Legal Diplomacy," Cornell Journal of International Law, No. 3 (Spring 1970), pp. 189-204.

37. HEENEY, A. D. P. and MERCHANT, L. T. Canada and The United States: Principles for Partnership. Ottawa: Queen's Printer (1969).

38.   CALDWELL, L. K. "Administrative Possibilities for Environmental Control" in DARLING, F. F. and MILTON, J. P. (eds.) Future Environments of North America. Garden City, New York: The Natural History Press (1966), p. 648.

39.   For example, see: JORDAN, F. J. E. "Recent Developments in International Environmental Pollution Control," McGill Law Journal, No. 15 (1969), pp. 279-301; REMPE, G. A., III. "International Air Pollution – United States and Canada – A Joint Approach," Arizona Law Review, No. 10 (Summer 1968), pp. 138-147; THOMPSON, A. R. "Policy Choices in Petroleum Leasing Legislation: Canada – Alaska Comparisons," in ROGERS, G. W. (ed.) Change in Alaska. College, Alaska: University of Alaska Press (1970), pp. 72-89; THOMPSON, A. R. "Basic Contrasts between Petroleum Land Policies of Canada and the United States," Colorado Law Review, No. 36 (1964), pp. 187-221; WANDESFORDE-SMITH, G. A Comparative Analysis of American and Canadian Governmental Arrangements for the Development of Regional Water Policy in the Columbia River Basin. Seattle: Department of Political Science, University of Washington, unpublished Ph. D. Dissertation (1971).

40.   Urbanization and Natural Environment, Report of a Conference held at the University of British Columbia, December 13 and 14, 1968. Vancouver: University of British Columbia, School of Community and Regional Planning and Seattle: University of Washington, Department of Urban Planning (April 1969), pp. 17, 40-41, 55.

41.   For the purpose of this study the Puget Sound and Strait of Georgia regions include all water and adjacent shorelines east of a line from Cape Flattery to Cape Beale, south of Johnstone Strait and north of Olympia.

CHAPTER 2
EXTERNALITIES, COMMON PROPERTY RESOURCES, AND INTERNATIONAL OIL POLLUTION

> Common property natural resources are free goods
> for the individual and scarce goods for society.
> Under unregulated private exploitation, they can
> yield no rent; that can be accomplished only by
> methods which make them private property or
> public (government) property in either case
> subject to a unified directing power.[1]

In Chapter One the basic problems of international oil pollution were discussed and reference was made to some of the peculiar problems associated with multiple uses of a common property resource such as the ocean. In this chapter these problems are defined and analyzed more rigorously within the conceptual framework of welfare economics and more particularly externality and common property resource theory. A specific attempt is made to correlate this analysis with the problems of international oil pollution and to suggest guides for evaluating and improving management programs at the state, provincial, national, and international levels.

Preventing and controlling oil pollution is only part of the total concept of water resource management and the ancillary problems of waste discharge and resource allocation. In the past economic values have frequently formed the bases for water resource allocation and environmental use decisions. This reliance on economic values cannot be rejected in any responsible approach to resource allocation, but it is necessary to place economic considerations in their proper perspective. Walter Firey's work, Man, Mind and Land – A Theory of Resource Use,[2] offers a tentative framework for a more realistic set of priorities in allocating resources. First, environmental decisions should be ecologically acceptable. Consideration of the ecological consequences of any decision is important insofar as it assures the permanence of the resource. Firey does not maintain

that there is universal agreement on what constitutes "ecological acceptabil-
ity" - he permits each society to define this in its own terms. Second, the
project should be socially desirable. Firey believes that any resource program must be particularly suited and adapted to the peculiar beliefs and
norms of the society in which it operates. Third, the project should be
economically feasible and efficiency criteria should be considered in allocating resources. Much of Firey's work was devoted to primitive societies, and extrapolation to more advanced societies has inherent dangers.
Prominent among these dangers is his neglect of advancing technology.
Often the available technology conflicts with the ecological and social
norms of the society. Modern technology permits offshore oil drilling on
the continental shelf, but oil spills off the coasts of California and Louisiana have prompted challenges to similar drilling in Puget Sound and the
Strait of Georgia. Firey's framework could be altered to include a technological component. Decisions on resource allocation would then have
to be: first, ecologically acceptable; second, socially desirable; third,
technically possible; and fourth, economically feasible. Given these priorities and constraints, the objective of any program seeking to prevent
international oil pollution is to optimize the allocation of ocean resources.
This will necessarily involve tradeoffs among the priorities, but the benefits from exploitation should be clearly shown to outweigh any damage that
might threaten the ecological stability of the region. The task of international water quality management is therefore, to determine the water quality necessary to provide optimum allocation for multiple use of the ocean,
to devise a physical system to achieve that quality, and to implement the
best institutional methods to manage the system.[3]

    While economic values must be placed in a broader decision-making
framework, the economic literature illuminates the difficulties of achieving
optimum use of ocean resources and provides guides for managing these resources. Much of this literature is concerned with the ability of a

competitive free enterprise market to deal effectively with problems of common property resources, externality, and environmental quality. To comprehend fully the problems associated with oil pollution of the oceans it is first necessary to outline the essential assumptions and tenets of a free market.

THE MARKET ECONOMY

In a model market economy, resources are hypothetically allocated through the competition of profit maximizing firms.[4] There is a decentralized decision-making system, the markets are competitive, and there is no domination by a few firms. Production activities are divisible or independent of each other in a real sense: that is, all inputs for a business (such as labor, raw materials, and machines) are under the control of the producer, and all products desired by the consumer are under his control.[5] The economy is not disturbed markedly by the introduction of new production methods or by changes in consumer habits. The wage for a given quality of labor is uniform in the market and there is universal agreement that the overall distribution of income and the market system is justifiable on ethical grounds.[6] Individual decision makers (individuals, firms, producers, and consumers) act rationally to maximize their private benefits, and individual preferences are satisfied to the maximum extent possible given the distribution of income.

Given these assumptions, production is organized to produce what each consumer desires within the limits of his income.[7] Each productive resource will be used up to the point where the cost of an additional unit is just equal to its contribution to the value of production. Firms will tend to produce items where the market price is equal to the opportunity cost, or to all alternative uses for the same amount of investment. Consumers will tend "to allocate their expenditures so that the last dollar spent for any particular item will yield an amount of satisfaction equal to the last dollar spent on any other item."[8] The results are threefold. First, the market will produce all the goods demanded by the consumer, produce them in just sufficient quantities and in the cheapest manner. Second, the market price of a good reflects the valuation that producers and consumers place upon the production and the purchase of marginal units. Third, there is an optimum allocation of resources and the welfare of producers and consumers is maximized, since all costs and benefits of all actions are indicated through the system of price and cost.[9]

MARKET FAILURE

In a utopian world the market economy would sustain economic activities and stop uneconomic activities. In the real world, however, many assumptions and tenets of a model free market economy are imperfect, and these imperfections result in a misallocation of resources. Bator[10] outlines the major imperfections in a utopian free market and suggests the market may fail for two reasons:

1. The existence of technological interdependencies. The production and consumption of one firm has a direct effect on another because there is a direct physical link between the output of one firm and the input of another. The market is often unable to cope with this situation because of the broad areal scope of the problem, the complexities of assigning damages where more than two firms are involved, and the absence of definable property rights. Technological interdependency implies the non-independence of various preference and production functions and results in a divergence between private and social cost-benefit calculations.[11]

2. The existence of public goods. In a sense this is a special case of indivisibilities in which it is not possible to define private ownership rights without markedly affecting the public welfare. An example of this could be a unique boating area where demand has not yet grown to the point where one person's right to utilize the waters has resulted in a marked reduction in another person's utilization of the same waters.

Wollman suggests that markets also fail when participants do not act in accordance with their own interests. We infer, he states:[12]

> that market action _per se_ is welfare maximizing, but this inference rests upon the prior assumption of adequate knowledge and foresight. To the extent that people act on the basis of restricted information and inadequate understanding of the consequences of their acts, market transactions fail to achieve maximum human welfare. Reliance on consumer sovereignty has its limitations.

This emphasis on consumer sovereignty is additionally important for the study area because the consumer is, at the same time, the people, the provincial and state governments, and the national governments of both Canada and the United States. Their interests, knowledge, and foresight differ considerably, yet all must be party to any international action in preventing oil pollution.[13]

The problem of preventing and controlling oil pollution is further compounded because each of the sources of market failure outlined by Bator would be present in the Puget Sound and Strait of Georgia regions if an oil spill were to occur. The water in the study area is the physical link that would carry any spilled oil and distribute it throughout the basin. Private property rights over salt water are ill-defined and practically non-existent. In territorial waters, both the United States and Canada have jurisdiction over any oil spillage, but can do relatively little to combat oil spillage outside their territorial waters or in the waters of the adjacent nation. The unique character and public nature of the waters in the study area are noted in Chapter Three. Any major oil spill would adversely affect the ecology and the recreational potential of these waters; yet, the market does not provide a structure for compensating those affected. In this case the waters are public or common property waters, not owned by or appropriated to any specific group. Thus, the market is irrelevant here unless private property rights can be established for salt waters.

EXTERNALITIES

This study is primarily concerned with market failure that results when producers and consumers do not have complete control over their respective inputs and consumption. That is, market prices are not the only forces governing production and consumption. The activities of one firm "may generate 'real', as contrasted with price or monetary effects, that are external to it. The economist refers to these as external effects or 'externalities'."[14] Externalities exist when:

$$c^A = c^A (X_1, X_2 \ldots X_m, Y_1)$$

where the consumption (c) of individual A is dependent on forces $(X_1, X_2 \ldots X_m)$ that are exclusively under his control or authority and forces Y_1 which are by definition, outside of his control and under that of individual B. In the present study, for example, an individual's use of beach front property may depend on his income, the facilities he owns near the property, and the time he wishes to devote to use of the beach. All these forces are under his control. Use of the beach may be impeded, however, by spilled oil over which the beachfront owner has no control. The spilled oil may cause the beachfront owner to stop using the beach or to incur certain expenses to clean up the beach and restore it to the original condition. Economists refer to these costs as "technological external diseconomies."[15]

The presence of external diseconomies does not necessarily disturb the equilibrium of market forces; however, they often conflict with society's notions of the optimum allocation of resources. To correct the discrepancy between the effects of external diseconomies and society's goals certain economic arrangements need to "be superimposed on those generated by the market."[16] These arrangements are becoming more important with the increased number of conflicts among individuals using a common property resource which is free but limited in supply.

Water is one resource where competition among individuals has not resulted in an optimum allocation of resources. Most of the documented problems of market failure in allocating water resources have concerned fresh water lakes and rivers, but the increased number of major oil pollution incidents has stimulated interest in limiting the effect of adverse externalities in salt water. Salt water bodies such as Puget Sound and the Strait of Georgia have traditionally had no price attached to their use. Thus pulp mills and domestic sewage systems have freely discharged wastes into the estuary. Governments have permitted such acts on the assumption that the water body is capable of absorbing the wastes without affecting subsequent users of the resource.[17] As a result excessive amounts of wastes are being discharged at several points throughout the estuary and these discharges are affecting the water quality of the whole estuary. External diseconomies resulting from industrial processes, domestic sewage systems, or spilled oil have become so pervasive as to make the market mechanism nonoperational.

The classical case of an externality in water resources is an upstream firm that is not forced to take account of the costs imposed upon downstream water users or the value of water use opportunities foreclosed by its effluent discharges. Any major oil spill in the Strait of Georgia or Puget Sound would probably present similar problems. A tanker mishap in the Gulf Islands, for example, would probably result in damage to the islands in the immediate vicinity and could cause damage in the adjacent San Juan Islands in the United States. The damages and costs resulting from such a spill would be external to the tanker polluting the waters. Under a free market the tanker could be regarded as underestimating pollution prevention costs in its output decision and creating a divergence between rational individual production or consumption decisions and welfare maximization. Resource allocation would not be optimized because of the inefficiencies that occur when costs are shifted from one independent economic unit to another. Kneese states the case explicitly:[18]

> ... when disposal of wastes into watercourses neglects
> downstream costs, the costs of some economic units are
> understated (apparently costless waste disposal into
> watercourses) and some are overstated ("excessive"
> imposed damages and treatment costs) relative to
> social (opportunity) costs. This tends to induce over-
> production and overconsumption of some items and
> underproduction and underconsumption of others.

To achieve a more optimum allocation of resources, it is necessary to a-
mend the operations of the market by devising a system of pollution control
and water treatment that identifies the contributions from specific polluters
and appropriate distribution of costs of pollution among economic units and
activities. If the costs of pollution were charged to the polluter, Kneese
argues that:[19]

> The firm [polluter] would equate the marginal costs of
> all relevant alternatives (water quality control measures
> and residual damages imposed by pollution), and the
> full marginal costs of each productive process would
> be considered in making output decisions. The re-
> sult of such a procedure would be "efficient" in the
> sense that the firm would produce the largest possible
> value for the resource it is using.

Thus, the critical problem is to develop a means whereby production decisions take into account all the costs and benefits resulting from such activity. If this can be accomplished the problem of misallocation can be
overcome.

COMMON PROPERTY RESOURCES

The rationale for public intervention in the market regarding water
resource management is provided by the externalities outlined above; however, any meaningful program to manage salt water resources requires an
understanding of the common property nature of the resource.[20] A consideration of common property resources can be approached in two ways.

The exploitation of such products as oil and fish represents extraction of resources from the environment. On the other hand, some industrial processes contribute to the deterioration of environmental quality by the addition of pollutants to the environment. These contrasting activities of extracting valuables from the environment and discharging destructive pollutants into the environment present similar management difficulties. Common property resources are shared by many individuals or nations; they are not legally alienated to individuals, firms, or nations for several reasons. Often a national desire to claim rights to a resource is complicated or deterred by international conventions designed to prevent individual ownership, by the mobility and fluidity of the resource, or by the anticipation of returns from exploitation that are lower than the social, political, and economic costs of alienating the rights.[21] Four fundamental problems arise when rights are not alienated and when "freedom in the commons" is condoned.[22] The resource tends to be used too quickly and is often depleted. Exploitation may be inefficient, employing more labor and capital than is required or justified by the economic rent that is derived. Frequently there is congestion among users of the resource.[23] Finally, the economic and legal systems have no way of accounting for externalities or extra party costs.[24] The critical difference between the two contrasting activities discussed above is that the extractive use, although at times uneconomical, adds directly to man's betterment, whereas the addition of pollutants contributes only to a decline in the quality of the environment and to an increase in treatment costs for subsequent users of the resource.

Gordon argues that exploitation of common property resources such as fish yields no economic rent and that the problems outlined in the previous paragraph are generally applicable in all cases where resources are owned by all and utilized under conditions of individualistic competition. He laments the attention given to maximization of catch by fishermen and the relative neglect of other factors of production which are used up in fishing,[25] although later works by Crutchfield and Pontecorvo and others[26] have partially rectified the dearth of knowledge about the economic costs involved in fishing. There is, as Gordon suggests, considerable similarity between the desires of fishermen to maximize their catch no matter what the production inputs of other factors and an almost universal desire to maximize utilization of the oceans. In the first case, we are extracting valuables from the environment, valuables whose contribution to society can be measured in terms of dollars or other comparable units. In the second case, we are using the ocean surface to transport oil and the ocean bed to drill for oil, activities which have increased the threat of serious ocean pollution. Both transportation and drilling are compatible with other uses of the ocean, and the value of these can be measured. When a major oil spill occurs, however, we are not always able to translate the value of recreational opportunities foregone, the impact on commercial enterprises, the region's reputation as a tourist area, or the ecological damage into measurable and/or comparable units. In the case of a major oil spill, utilization of the oceans has the classic characteristics of "overuse" (or overfishing) as Gordon argues, but it also has deeper implications that are apparent as one analyzes the tenets of Gordon's theory in light of a foreign material being added to the ocean.

Gordon develops his theory around a typical demersal fish such as cod, but argues that the conclusions he makes would be applicable to exploitation of any other fish. He assumes that fish live in relatively specified zones, that total cost and production can be expressed as a function or degree of fishing intensity, that as fishing effort expands the catch of fish increases at a diminishing rate, and that the functional relationship between average production (production per-unit-of-fishing-effort) and the quantity of fishing effort is uniformly linear. Gordon concludes that the long-run maximum sustained physical yield will be attained when the marginal productivity of fishing effort is zero and that the optimum economic

fishing intensity (that point at which economic rent is greatest) is less than
that which would produce the maximum sustained physical yield. The real
problem is that the rent the resource might yield cannot be appropriated to
anyone so that, given two fishing grounds of differing fertility and/or location, fishermen will gravitate towards that ground where average productivity and total yield are greatest. An optimum distribution of fishermen
would result when the marginal productivities are equal on both grounds,
but fishermen are unwilling to abide by such an allocation when they might
be able to secure greater yields on one ground. The result is overuse of
fishing areas, decreasing stocks, and restrictive regulations which make
fishing economically inefficient.[27]

Gordon's conclusions about economic efficiency and overuse are
applicable to ocean pollution; however, the differences in production
functions, the mobility of the resource, the nature of the pollutant, the
social consequences of ocean pollution, and the problem of jurisdiction
over the resource must also be considered in any management program for
the oceans. Total cost and total production emanating from use of the o-
cean can be expressed as a function of the number and degree of uses. We
cannot assume, as Gordon does, that the functional relationship between
average production is uniformly linear. Such a relationship may be conceptually convenient but there is not a linear function between the amount
of pollution and the number of polluting units (see Figure 1). Most
pollution concentrations have nonlinear physical consequences and the
marginal cost of alternative treatment techniques is not constant over the
full range of desired waste disposal concentrations.[28] We can therefore
assume that: (1) as the number of uses of coastal waters increase, and
(2) if the amount of localized pollution intensifies, and (3) if waste discharges are not controlled, then the threat posed by a major pollutant
such as an oil spill will be more critical than if a spill of the same magnitude occurred in relatively pollution free waters. Under these conditions,

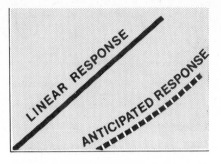

Concentration

Water Quality

Waste Discharged ⟶

FIGURE 1   Relationship between
amount of pollution and number of polluting units.

Source: Gardner Brown Jr. and Brian Mar, "Dynamic Economic
Efficiency of Water Quality Standards or Charges, Water
Resources Research, vol. 4 (December, 1968), p. 1154

water quality in a basin would not be greatly affected by an increase in
localized pollution because of the capacity of the receiving body to dissipate the pollution, but would deteriorate rapidly in the case of a major
oil spill.

When oil is spilled on salt water, it takes on all the properties of
a common property resource. Oil flows with the currents and tides in
coastal waters. In the study area, the most critical international problems
would likely emanate from areas where the mobility of the water and oil
is greatest; that is, in the immediate waters along the outer coast, the
Strait of Juan de Fuca, and the San Juan and Gulf Islands.

Gordon assumed that the mobility of a common property resource
would not significantly alter his theory; however, oil spilled in the territorial waters of country "A" that inflicts damages on country "B" can cause
immediate economic costs and political confrontation. Similarly a tanker
registered in country "C" could spill oil on the high seas and could inflict
damage on both countries "A" and "B". In either case, where spilled oil

PLATE 5
Monitoring oil damage from the Vanlene grounding, Vancouver Island

Photo: W.C. Austin

flows with the currents and affects two or more nations an externality is
created, although this particular occurrence would be a special international case of externality. Gordon's model fails to incorporate costs which
would be imposed by the mobility of this oil and therefore the externality,
nor does he give sufficient attention to the social and political consequences
resulting from a major oil spill in the waters described above.

We cannot assume that efficiency criteria will lead to a social and
political optimization of resource allocation. Public response to major
oil spills in the United Kingdom, California, Louisiana, and Nova Scotia
suggests that governments are under intense pressure to clean up spills and
return the water to a quality at least equal to that prior to the spill. Governments have been largely unable to solve immediate problems because
they lack the technology and trained forces to deal with a major spill.
Some governments have not had adequate powers to prevent oil spills, while
others have not formulated plans to deal with such disasters. As a result,
governments are under continuing pressure to introduce measures which will
protect and aid those affected by oil spills. The results may not lead to
greater economic efficiency if costs imposed on polluters exceed benefits,
or if the governments are not able to charge the polluters; but they can be
politically expedient and possibly a more socially desirable allocation of
resources.

Gordon postulates that lack of appropriation causes inefficiencies
in resource allocation, but fails to explicitly recognize that lack of appropriation may be due to an absence of jurisdiction over the resource. Lack
of jurisdiction can be critical in preventing and controlling international
oil pollution. Oil spilled on the high seas or in the territorial waters of a
nation state is not appropriated to any nation; yet, if that same oil fouled
the shores of another nation state, that state must clean up the spill. International conventions recognize an obligation on the part of the polluter, but contain no compulsory measures making the polluter legally liable

46

for the pollution.[29] When the polluter fails to incorporate the costs caused by the pollution there can be a misallocation of resources.

Gordon analyses the classic case of overuse when goods are extracted from the commons but does not incorporate some of the real problems associated with adding pollutants to the commons into his model. The result is that Gordon places a great emphasis on economic efficiency in allocating resources and neglects many of the ecological, social, and political problems emanating from pollution of the environment. The lack of jurisdiction, the political and social consequences, the mobility, and differences in production functions associated with some uses of the commons must be considered along with the efficiency criteria in any program seeking to minimize the adverse effects of externalities created by an international oil spill.

REMOVING EXTERNALITIES

Limiting the adverse effect of externalities can lead to a better allocation of resources. Inherent in all alternatives to limit the adverse effects of spilled oil is a recognition that some property rights must be applied to ocean resources held in common. Dales states the case for this need when he argues that:[30]

> Unrestricted common property rights are bound to
> lead to all sorts of social, political, and economic
> friction, especially as population pressure increases,
> because, in the nature of the case, individuals have
> no legal rights with respect to the property when
> its government owner follows a policy of "anything
> goes." Notice, too, that such a policy, though
> apparently neutral as between conflicting interests,
> in fact always favours one party against the other.
> Technologically, swimmers cannot harm the polluters,
> but the polluters can harm the swimmers; when
> property rights are undefined those who wish to
> use the property in ways that deteriorate it will
> inevitably triumph every time over those who wish

47

> to use it in ways that do not deteriorate it.
> Economically and socially the question is always
> which set of interests should prevail, or rather
> what sort of accommodations should be made among
> the various interests concerned. The question is
> always, and inescapably, the great question of
> social justice.

Some of the simplistic alternatives can be rejected as inapplicable to international oil pollution. Private agreements between the polluter and
the activity being polluted about acceptable levels and possible compensation are not a likely solution because: (1) there would be an excessive
number of people to be compensated if oil were spilled; (2) the threat
posed by an oil spill is not a continuous hazard; and (3) binding international agreements between individuals and firms are cumbersome and difficult to administer and enforce. Legal action through the courts is a slow
and tedious procedure. If the oil spill affected only domestic waters, legal processes could possibly hold the polluter liable for damages in the region. Kneese and Bower have examined the use of legal procedures and
conclude that court costs, variability and dispersion of damages, and definition of those polluted are cumbersome and lead to an inefficient allocation of resources.[31] When an oil spill has international dimensions both
parties to the spill have to agree to submit their case to a court and abide
by the court's decision. This creates, in addition to the constraints imposed by domestic courts, all the problems inherent in a private agreement
between the polluter and those polluted.

Dales suggests that forcing the polluter to internalize externalities
is the most efficient way to correct misallocation of resources.[32] Alternatively, many economists hold that technological spillovers can be counter-
acted by levying taxes on the unit "responsible" for the diseconomy and by
paying a subsidy to the damaged party.[33] The goal of such a program
would be to minimize the overall costs associated with waste disposal in
the region. If costs are properly defined, and if true costs are charged to

the waste discharger, the result will tend towards optimality for both water quality and waste loads. While most social scientists agree on the goals, many disagree on a feasible technique. Effluent standards, payments to reduce discharge, and charges on effluents have all been suggested as a means of overcoming externalities. Each technique must satisfy several requirements if it is to prevent and control international oil pollution. First, the technique must be capable of operating within the framework for optimizing allocation of ocean resources suggested earlier in this chapter. Second, any agency or government body managing the area must have sufficient spatial jurisdiction to internalize the externalities. Third, the agency or body must have sufficient political support and be able to define and enforce property rights. Fourth, it must be capable of reacting quickly to control major spills before the oil spreads.

Standards, defined in terms of the amount of waste discharged, could be used by a regional agency to achieve optimum resource allocation; however, they are not particularly suited to a problem of ocean pollution. Standards are most appropriate where there is a continuous flow of pollutants at one specific location. When the pollutant flows continuously a quality standard can be set so that incremental changes in the standards will equate marginal treatment costs with the marginal costs of the damages avoided.[34] The imposition of effluent standards, however, provides no incentive to curb waste discharge beyond the required level even though it may be possible to do so quite inexpensively, and implies that beyond a certain point pollution is costless. Standards depend on precise engineering, ecological, and economic information that is expensive and difficult to obtain. In this sense they are no different from charges and payments described below. The critical difference among standards, charges, and payments rests upon efficiencies in administration and concepts of equity. When pollution is not continuous and not fixed at one location, additional problems arise. Difficulties emerge in trying to predict ecologically safe standards for a disastrous and sudden oil spill. In the study region ecological characteristics of various areas differ, and one uniform water quality standard in such a small area would be complex and costly. Moreover, further costs would be incurred (presumably at government expense) in response to cleanups associated with major spills.

Under a system of charges, each polluter is charged a fee per unit of waste discharge equivalent to costs resulting from this discharge of waste. Conceptually a payment is the converse of a charge and each polluter is paid a fee to desist from discharging wastes whose cost equivalent is equal to the fee. Charges are not perfect but they are preferable over both standards and payments in that they can approximate the requirements that are necessary to prevent and control oil pollution. A charge offers incentives for each firm to take action to the lowest level of waste discharge whether through preventive techniques or insurance against oil spills. Charges do not have the same degree of misallocation effects that may accompany standards and payments because charges tend towards an equalization of marginal costs such that the incremental costs of waste reduction are kept in balance with the charge. Effluent charges yield revenues to some extent whereas standards and payments do not. A portion of those revenues could be used to finance a program with trained personnel ready to respond to any major oil spill. Charges have additional advantages over payments because they are more likely to agree with the public's concept of justice, are easier to administer, and are not as susceptible to extortion.[35]

Charges appear to be the best means of achieving a more efficient allocation of resources provided that the priorities for resource allocation, the real costs of waste discharge, and the requirements for preventing and controlling international oil pollution can be incorporated substantially within a set of charges. Kneese and Bower argue that it is possible to incorporate many of these costs within a set of charges by basing the charge

on the damage function (functional relationship between the amount of
waste discharged and the damage caused). When damages are linear a
level of charges equal to the incremental damage cost can be worked out.
Firms could then respond rationally to imposed charges by assessing benefits and total costs of their production decisions. They further argue that
it is better to assume linearity in the discharge-damage relationship even
when it is non-linear because the attempt to take account of greater complexity may rapidly increase costs and yield quickly diminishing returns.
The costs of refinement of a system of charges must not be neglected.
Kneese and Bower place great emphasis on optimizing economic resource
allocation for present conditions.[36]

Brown and Mar take a broader view of resource allocation arguing
that if suitable economic and physical conditions can be defined for one
time period it is possible to define an optimum water quality over time and
establish appropriate water management charges. They recognize, as noted
above, that the damage function is not linear and that a non-linear damage
function must be incorporated within any system of charges. Moreover,
Brown and Mar argue that any set of standards or charges must be placed
into a more dynamic perspective. Because various lags[37] affect water
quality management they question:[38]

> ... whether the attempt to establish water quality at
> its optimum level at every point in time is superior
> to a "farsighted" rule that sets stream quality at
> a level that is suboptimum today but that will
> become and remain optimum when the economy
> reaches a steady state of growth.

By incorporating an ecological component within an economic charge
Brown and Mar are better able to approximate the full social costs of re-
ductions in water quality. Parker and Crutchfield[39] demonstrate this point
well. They argue that pollution of water by one user often precludes use
by an alternative activity that would have shown a higher growth rate over

51

time:[40]

> This results in a systematic and disturbingly large under-
> statement of the real costs of water pollution. Since
> many of the resulting planning errors are irreversible,
> in an economic if not in a physical sense, the resulting
> malallocation of resources is further accentuated.

Barring some new production process, the most efficient technique by which oil pollution can be controlled is by forcing the polluter to internalize the technological externality[41] created by his waste discharge through a system of effluent charges based on a non-linear damage function. These charges will allow him to respond more rationally at several points in time in his output decision by knowing all costs and benefits that will accrue from each decision. Assessing abatement and damage costs as reflected by the charge he will move toward a point of social desirability and economic efficiency in his output provided that the charge exceeds the abatement cost. Under these conditions the costs of waste disposal will be minimized.

The system of charges described above incorporates many of the priorities for resource allocation and provides a system to implement desired water quality standards, but fails to consider some of the critical political and administrative difficulties in implementing a program to prevent and control international oil pollution. If the quality of the ocean environment is to be improved, international organizations and nation states must formulate aims for management programs, delineate those responsible for overuse or misuse of the ocean, select the individuals or firms to be restricted, and develop the means of restriction and enforcement. The chapters which follow examine, within the context of guides suggested in this chapter, the national and international programs seeking to prevent and control oil pollution in the Puget Sound and Strait of Georgia regions of Washington and British Columbia.

REFERENCES

1. GORDON, H. "The Economic Theory of a Common Property Resource: The Fishery," _Journal of Political Economy_, No. 62 (April 1954), p. 135.

2. FIREY, W. Man, Mind and Land - A Theory of Resource Use. Glencoe: The Free Press of Glencoe, Illinois (1960); see also CASTLE, E. N. and STOEVENER, H. H. "Water Resources Allocation, Extra-market Values and Market Criteria: A Suggested Approach," _Natural Resources Journal_, No. 10 (July 1970), pp. 532-544. Castle and Stoevener make the same point although they place a strong emphasis on the role of the market in allocating scarce water resources. "Even though the market is rejected [in their article] as a means of allocating certain goods and services, it may still provide data and criteria in dealing with extra-market problems. The role of the market in generating relevant information for decision-makers has not been given the emphasis it deserves... We do not argue for complete dominance of the market; we do argue for the kind of rationality that market logic can bring to social decision-making." (p. 544).

3. KNEESE, A. V. The Economics of Regional Water Quality Management. Baltimore: Johns Hopkins Press (1964), pp. 191-206; KNEESE, A. V. and BOWER, B. T. Managing Water Quality: Economics Technology, Institutions. Baltimore: Johns Hopkins Press (1968), pp. 5-7.

4. The fundamental elements of a market economy are outlined in many books and articles. A simple, yet comprehensive presentation can be found in DOOLEY, P. C. _Elementary Price Theory_. New York: Appleton-Century-Crofts (1967).

5. HERFINDAHL, O. C. and KNEESE, A. V. Quality of the Environment: An Economic Approach to Some Problems in Using Land, Water and Air. Baltimore: Johns Hopkins Press (1965), p. 5.

6. KNEESE, op. cit., pp. 38-39.

8.   KNEESE and BOWER, op. cit., p. 76.

9.   Ibid., pp. 75-78.

10.   BATOR, F. M. "The Anatomy of Market Failure," Quarterly
        Journal of Economics, No. 72 (August 1958), pp. 351-
        379; see also KAPP, K. W., The Social Costs of Pri-
        vate Enterprise. New York: Schocken (1970).

11.   The divergence between private and social cost-benefit calcula-
        tions is dealt with more fully in COASE, R. H. "The
        Problem of Social Cost," Journal of Law and Economics,
        No. 3 (October 1960), pp. 1-44; MEADE, J. E.
        "External Economies and Diseconomies in a Competitive
        Situation," Economic Journal, No. 62 (March 1952),
        pp. 54-67; WILLIAMS, B. R. "Economics in Unwonted
        Places," Economic Journal, No. 75 (March 1965),
        pp. 20-30.

12.   WOLLMAN, N. "The New Economics of Resources," Daedalus,
        No. 96 (Fall 1967), p. 1101.

13.   The problem of defining and dealing with consumer interests in in-
        ternational resource management is clearly demonstrated
        in the prolonged negotiations over the Columbia River
        during the 1950's and 1960's and the controversy over
        Seattle City Light's decision to raise the level of Ross
        Dam. Paddy Sherman offers an illuminating study on
        the differences over the Columbia in his book Bennett.
        See SHERMAN, P. Bennett. Toronto: McClelland
        and Stewart (1966). The Ross Dam problem stems from
        a decision by Seattle City Light to raise their dam on
        Ross Lake to provide additional hydroelectric power for
        Seattle. This decision is related to an agreement made
        with the British Columbia government to flood a portion
        of the adjacent Skagit Valley in British Columbia. Lo-
        cal preservationists in British Columbia, outraged at the
        concurrence of the British Columbia government, have
        attempted to persuade the Canadian federal government
        to disallow the agreement. Both problems stem largely
        from a failure to clearly identify consumers, establish
        their needs, and weigh their political power.

14.   See KNEESE and BOWER, op. cit., p. 77 for a summary of

externalities. Externalities have also been referred to
as "spillovers," "external economies and diseconomies,"
"side-use effects," "off-site costs," "extra party costs,"
and "third party effects".

15. MEADE, op. cit., pp. 54-67. See also SCITOVSKY, T.
     "Two Concepts of External Economies," Journal of
     Political Economy, No. 62 (April 1954), pp. 143-151
     and DAVIS, O. A. and WHINSTON, A. "Externali-
     ties, Welfare and the Theory of Games," Journal of
     Political Economy, No. 70 (June 1962), pp. 241-262.

16. MISHAN, E. J. "Reflections on Recent Developments in the
     Concept of External Effects," Canadian Journal of
     Economics and Political Science, No. 31 (February
     1965), p. 5.

17. KNEESE and BOWER, op. cit., p. 80.

18. KNEESE, op. cit., p. 42.

19. KNEESE, A. V. Water Pollution: Economic Aspects and Research
     Needs. Baltimore: Johns Hopkins Press (1962), p. 29.

20. Barbara Ward has argued that all resources, even those of the
     nation-state should be regarded as part of a common re-
     source of this "spaceship earth". See WARD, B. Space-
     ship Earth. New York: Columbia University Press (1966)
     and BOULDING, K. E. "The Economics of the Coming
     Spaceship Earth" in JARRETT, H. (ed.) Environmental
     Quality in a Growing Economy. Baltimore: Johns Hop-
     kins Press (1966), pp. 3-14. In this study, common pro-
     perty resources are those resources, such as salt water,
     where there are few defined property rights or where
     the fluidity and mobility of the resource bring it under
     the jurisdiction and control of two or more nation-states.
     Compare these definitions with J. H. Dales' analysis of
     unrestricted common property resources. See DALES,
     J. H. Pollution, Property and Prices. Toronto: Univer-
     sity of Toronto Press (1968), pp. 58-76.

21. CHRISTY, F. T., Jr. "Efficiency in the Use of Marine Resources,"
     Resources for the Future, Reprint 49 (September 1964),
     pp. 1-8.

22. HARDIN, G. "The Tragedy of the Commons," Science, No. 162
          (December 13, 1968), pp. 1243-1248.

23. CHRISTY, op. cit.; GORDON, op cit., pp. 124-142.

24. WOLLMAN, op. cit., pp. 1099-1104.

25. GORDON, op. cit., pp. 124, 128; SCOTT, H. S. "An Econ-
          omic Approach to the Optimum Utilization of Fishery
          Resources," Journal of the Fisheries Research Board of
          Canada, No. 10 (October 1953), pp. 442-457.

26. CRUTCHFIELD, J. A. and PONTECORVO, G. The Pacific Sal-
          mon Fisheries: A Study in Irrational Conservation.
          Baltimore: Johns Hopkins Press (1969); CHRISTY, F. T.,
          Jr. and SCOTT, A. The Commonwealth in Ocean
          Fisheries: Some Problems of Growth and Economic Al-
          location. Baltimore: Johns Hopkins Press (1966).

27. GORDON, 1954, op. cit., pp. 128-135; See also CRUTCHFIELD,
          A. "An Economic Evaluation of Alternative Methods
          of Fishery Regulation," Journal of Law and Economics,
          No. 4 (October 1961), pp. 131-143.

28. BROWN, G., Jr. and MAR, B. "Dynamic Economic Efficiency
          of Water Quality Standards or Charges," Water Resources
          Research, No. 4 (December 1968), pp. 1153-1159.

29. O'CONNELL, D. M. "Reflections on Brussels: IMCO and the
          1969 Pollution Conventions," Cornell Journal of Inter-
          national Law, No. 3 (Spring 1970), pp. 161-188.

30. DALES, op. cit. pp. 67-68.

31. KNEESE, A. V. and BOWER, B. T. Managing Water Quality:
          Economics Technology, Institutions. Baltimore: Johns
          Hopkins Press (1968), pp. 83-89.

32. DALES, op. cit. pp. 77-100.

33. KNEESE, A. V. The Economics of Regional Water Quality Man-
          agement. Baltimore: Johns Hopkins Press (1964),
          p. 56.

34.   CASTLE, E. N. and STOEVENER, H. H. op. cit., p. 540.

35.   KNEESE and BOWER, op. cit., pp. 136-140.

36.   Ibid.

37.   Brown and Mar outline five possible lags with which a water qual-
      ity manager must cope in a dynamic setting: (1) the in-
      formation lag occurs between the time a change in the
      environment takes place and the time information be-
      comes available about the change; (2) the recognition
      lag intervenes between the arrival of information and
      the recognition of the need for action; (3) the evalua-
      tion lag refers to the time it takes to evaluate the
      changed circumstances and to decide on a suitable sub-
      stitute course of action; (4) the application lag is the
      time needed to implement a change in policy; and (5)
      a lag may occur between the action and its consequences.
      See BROWN and MAR, op. cit., p. 1158.

38.   Ibid., p. 1159.

39.   PARKER, D.S. and CRUTCHFIELD, J.A. "Water Quality Man-
      agement and the Time Profile of Benefits and Costs,"
      Water Resources Research, No. 4 (April 1968), pp.
      233-246.

40.   Ibid., p. 234.

41.   McGURREN, H.J. "The Externalities of a Torrey Canyon Situation;
      An Impetus for Change in Legislation," Natural Re-
      sources Journal, No. 11 (April 1971), pp. 349-372.

CHAPTER 3

THE THREAT OF AND RESPONSE TO OIL POLLUTION IN PUGET SOUND AND THE STRAIT OF GEORGIA

Chapter Three consists of four parts. Part one examines the susceptibility of the region to oil pollution and defines the perimeters for examining domestic legislation and international treaties as they apply to Puget Sound and the Strait of Georgia. Parts two and three respectively outline the response of the United States and Washington, and Canada and British Columbia to the problem of international oil pollution in the study area. Part four evaluates the response of the national and subnational governments to the international threat.

THE THREAT OF OIL POLLUTION IN PUGET SOUND AND THE STRAIT OF GEORGIA

Puget Sound and the Strait of Georgia are essential components of an immense salt water inland estuary, which is fed by the Fraser, Skagit, Snohomish, Nisqually, and other watersheds, and has the combined and interrelated characteristics of a bay, estuary, and coastal region (see Figure 2). In many parts of the region water transfer is slow, although in the two entrances to the estuary, the Strait of Juan de Fuca and Johnstone Strait , tidal currents and winds result in considerable mixing and transfer. While water is transferred out as in other estuaries, the study region poses unique problems for water resource management because it is a salt water body and is surrounded by two countries, one province, one state, and several cities, municipalities, and counties. Most management agencies, both national and international, have been concerned with fresh water and have organized themselves on the river basin concept. These particular management organizations are not appropriate for a salt water estuary since this type of body lacks the physi-

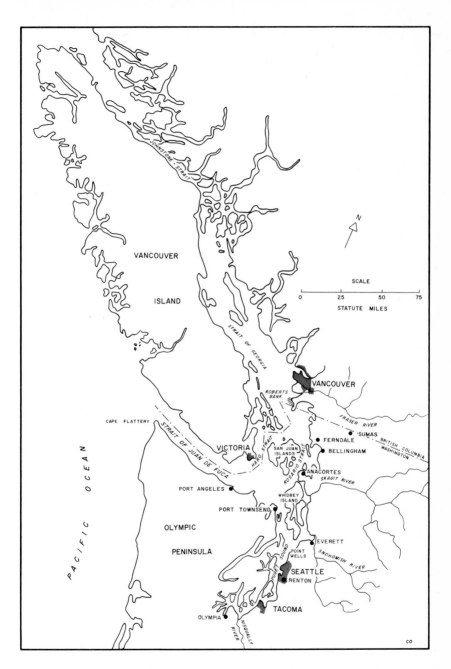

FIGURE 2   The Puget Sound and Strait of Georgia Regions

cal and ecological characteristics common to most river basin organizations. In addition, much less is known about the estuarine dynamics of
Puget Sound and the Strait of Georgia beyond highly specialized surveys
for particular localities. Many of these surveys have been made in areas
where pollution from untreated sewage and industrial processes have adversely affected water quality. The effects of these pollution incidents
have been highly localized, present no immediate danger to the ecological balance in Puget Sound and the Strait of Georgia, and are probably
easiest to control and regulate through domestic legislation.

Oil spillage in the study area is potentially more threatening and
dangerous to the water quality of the estuary than localized pollution.
Oil is not absorbed well by salt water, and natural degradation is dependent on satisfactory environmental conditions (nutrients, sunlight, tem-
perature, and oxygen availability) and suitable microbial populations
which can absorb the oil. Even with satisfactory conditions natural degradation is slow.[1]

If oil were spilled near the international boundary, it would not
necessarily respect the artificial political divisions and remain on the side
in which it was spilled. Oil that is spilled and affects only domestic waters is largely beyond the scope of this study. The study is concerned only with international oil spills — an international spill being defined here
as any commercial vessel discharging oil along or near the international
boundary and affecting the waters of both Washington and British Columbia. National and subnational governments do not own the oceans, but
they must contend with hazardous materials that are transported and
spilled on the ocean.

The greatest threat of international oil pollution stems from two
sources. The first of these is the transport or highly toxic refined products
within the estuary. Most of the barges and tankers participating in this
trade are small, but the number of sailings and volume of oil transported

is considerable. The second of the threatened sources of pollution is oil and refined products that are entering or leaving the region by large tankers. Nearly all oil coming and leaving the region must go through the Strait of Juan de Fuca. The concentration of tanker and barge traffic along with other vessels in the area of the international boundary increases the probability of an oil spill. If oil were spilled in this area, it could move quickly and foul beaches and water in the San Juan and Gulf Islands, spreading damage over a much larger area than could small a-mounts of localized pollution. Efforts to prevent and control spills and institute compensation for those damaged from oil pollution in Puget Sound and the Strait of Georgia are complicated by the international boundary which bisects the natural entrance to and physical integrity of the region.

Susceptibility of the Region to International Oil Pollution

Puget Sound and the Strait of Georgia are as susceptible to oil pollution as other regions discussed earlier in Chapter One. The region is unique, however, in four major aspects. First, the water circulation, the tides, and the winds could in "malimum" circumstances, cause oil spilled in Puget Sound to spread over all of Georgia Strait.[2] The vigorous mixing of water, especially in the area of the international boundary, heightens the possibility of international complications from an oil spill on either side of the boundary. Second, the heavily indented coastline and the multitude of islands opens hundreds of miles of beaches to damage from oil pollution. Third, the region is a major resting station on the Pacific flyway. Marshland areas in the region would be particularly vul-nerable to oil pollution during migration periods. Fourth, Puget Sound and the Strait of Georgia have valuable commercial fisheries which are particularly susceptible to oil at different seasons. Some of the valuable salmon runs entering the region, for example, could be severely depleted if a major oil spill occurred in the Strait of Juan de Fuca between June and September.[3] There are numerous studies describing the water circulation in particular areas and we know the wildlife species that migrate through the region. But there is no atlas or other compendium summarizing available knowledge concerning seasonal composition of species and water and wind movements and correlating this information with the possible ecological effects from various kinds of petroleum products or from methods proposed for spill control. Without this knowledge efforts to control spills remain crisis oriented. The existence of contingency plans based on incomplete knowledge increases the susceptibility of a region to damage from oil pollution.

Current State of the Oil Industry

Demand for Oil

The oil industry in the study area has grown primarily in response to local demands and the region is not a major transhipment center for either crude oil or refined products. Prior to 1954 all crude oil refineries within the region had to be imported either by tanker or rail as no pipelines existed to transport crude oil from interior fields. Most of this oil came from California but the amount imported was small because of the limited market. With the discovery of major oil deposits in Alberta and a rising demand in the study area, Trans Mountain Oil Pipe Line Company constructed a pipeline from Edmonton to Vancouver (see Figure 3). In addition a spur line was laid to the international boundary near Sumas and thence through a line owned by Trans Mountain Oil Pipe Line Corporation to Mobil Oil's Ferndale refinery in 1954, Shell Oil's Anacortes refinery in 1956, and Texaco's Anacortes refinery in 1958. Initial designs were based on a throughput of 200,000 barrels per day (b/d) but, as new refineries were built, capacity was increased to a throughput of

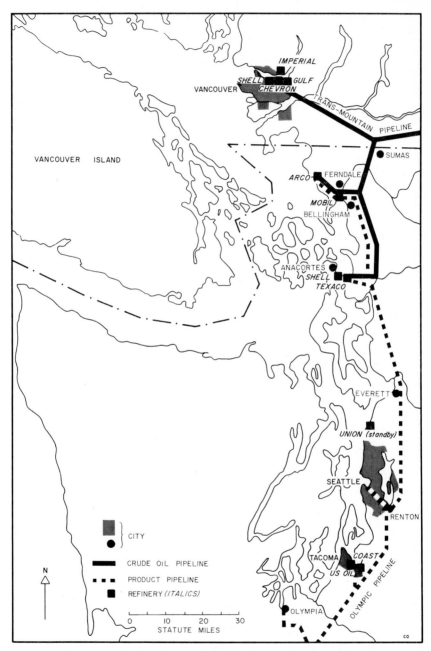

FIGURE 3   Major Oil Pipelines and Refineries on Puget Sound
and Strait of Georgia

315,000 b/d and in mid-1971 reached a capacity of 400,000 b/d. The line now serves five refineries in British Columbia with a capacity of 103,900 b/d and three refineries in Washington with capacities of 189,000 b/d. Plans have been made to connect the new Atlantic Richfield refinery at Cherry Point to the line and Alberta crude will supply the initial needs of this refinery.[4] Even with construction of the Cherry Point Refinery the Trans Mountain Pipe Line is capable of meeting the requirements of all these refineries, given no restriction on the importation of Canadian crude into the United States.

Oil Flows in the Region

Small amounts of crude are imported by tanker to refineries at Ferndale and Anacortes and other refineries in Washington at Point Wells and Tacoma, but the amount is small in comparison with pipeline deliveries. Oil has been exported from Trans Mountain's deepsea berth in Burrard Inlet, but until 1970 the shipments were sporadic. With the prevailing freight rates on tanker shipments from the Middle East to the United States, it has become economical to ship Alberta crude from Vancouver to markets in California. This movement is expected to continue at the rate of a couple of sailings per month totalling between 300,000 and 400,000 barrels of crude.[5]

The flow of refined products within the region is much more complicated and potentially more threatening to the marine environment because of the products' toxicity and the method of transport used. Little data is available on the amount of oil moving within the region. Data that are available on ship movements can be very misleading since empty barges or tankers are recorded in the same manner as those carrying products, but do not pose as great a threat to the environment. Approximately 26 percent of the oil refined in northwest Washington is further transported through the Olympic Pipe Line, which was completed in 1964 and extends south as far as Portland, Oregon via a route inland east of Bellevue to Harbor Island (Seattle), Tacoma, Olympia, and Portland. Approximately 69 percent of the refined oil is transported by barge with an estimated seven barges traversing Puget Sound every day, carrying 220,000 barrels of refined products.[6] While Washington meets the majority of its own needs (as well as some of the demand in Idaho and Oregon) through its own refineries, some refined products are imported to supplement local production. In addition, many of the pulp mills in the state, as those in British Columbia, import substantial amounts of bunker fuel for plant operations, mainly from California. Each mill, as well as some other industries, can use between 100,000 and 300,000 barrels per year. The refining capacity in British Columbia is smaller than that of Washington; however, more products move by water since the only pipe lines carrying refined products are close to the refineries and service such facilities as Vancouver International Airport. All products destined for Vancouver Island and the coastal reaches of northern British Columbia must be transported at least partially by water. The various oil and towing companies operate 22 barges and 4 tankers with a capacity of 9,560,000 gallons.[7] Not all of these operate at any one time within the Strait of Georgia; however, the great bulk of the coastwise petroleum trade is conducted south of Johnstone Straits. Any increase in the number and size of tankers and barges currently operating within the study area will increase the potential damage from spilled oil.

Offshore Drilling

Offshore drilling currently poses no threat to the marine environment because governments have become increasingly reluctant to issue permits for exploration and drilling in the absence of an immediate need for oil in the study area. In Washington State, oil companies applied to the State Land Commissioner for permission to explore and drill in Puget Sound.

Hearings were held and many preservation groups such as the Sierra Club, the Audubon Society, and the Washington Environmental Council, and some counties bordering on the Sound vigorously opposed the applications, arguing that the revenues would not be sufficient to compensate for potential damage. On November 16, 1970, Bert Cole, the State Land Commissioner, refused to issue permits to the oil companies but intimated that he was prepared to reverse his decision at some later date under specified conditions. First, it must be clearly demonstrated that oil from the Sound would significantly benefit the state's economy. Second, the oil companies would have to ensure that stringent safety devices would reduce the number of potential spills, that effective plans and equipment would be available to minimize the effect of any spill, and that a program would be instituted to compensate those damaged by any spill in accordance with state law.[8] The situation in British Columbia is complicated by a jurisdictional dispute over who has the right to issue permits for exploration and drilling in offshore waters. In the early 1960's both the federal and provincial governments were issuing permits for seismic exploration off the British Columbia coast and both claimed jurisdiction over all offshore minerals.[9] Several companies including Gulf Oil and Texaco carried out seismic work under permits from both governments. Shell Oil did some drilling but discovered no oil. To resolve the dispute over offshore mineral rights, the British Columbia and Canadian governments agreed to submit the issue to the Supreme Court. In a 1967 decision the Supreme Court ruled that the Canadian government has sole rights to all subsurface minerals below the low water line. The provincial government refused to regard the decision as final and stated that the decision would be merely an aid in political negotiations between the two governments. The federal government, while offering concessions immediately after the Supreme Court decision, hardened its attitude after the election of the Trudeau government and the appointment of Jack Davis as Minister of Fisheries and

Forestry in 1968.[10] Davis stated there would be no drilling in the Strait of Georgia. He maintained that the Strait:

> is obviously a priceless asset from a recreation point of view and it's a funnel through which a 100 million dollar fishery moves. The benefits [of drilling] are just not there. There is no way - if you blew up the price and employment benefits a hundred times they would be there.

As a result oil companies asked the provincial government to suspend their leases. Ottawa requested and the companies agreed to surrender federal permits. The federal government then announced that no further exploration for oil or natural gas would be undertaken.[11] Until and unless the federal-provincial dispute is resolved, exploration and drilling in the Strait appears to have been curtailed.

Previous Oil Spills

The transport of refined and crude oil within Puget Sound and the Strait of Georgia has not resulted in any catastrophic spills, but several incidents have demonstrated that a threat does exist. Most of the recorded oil spills have resulted from bilge pumping and barge operations.[12] There have been a number of incidents where the fuel tanks on commercial vessels have been punctured and caused localized damage. In Washington, a small spill of 1,600 gallons of bunker oil near Bellingham spread rapidly and nearly four days elapsed before it was cleaned up. On April 26, 1971, 210,000 gallons of diesel fuel was spilled near Anacortes and killed birds and other wildlife. It also damaged beaches in Washington and threatened beaches near Victoria. In British Columbia, a fuel barge carrying 300,000 gallons sank in Howe Sound and 100,000 gallons spread from the barge before it could be raised.[13] None of these incidents caused damage to the extent seen in Nova Scotia or in California, but they do focus attention on the possibility of a major spill.

Future Growth of the Oil Industry

Offshore Drilling

The threat of a major oil pollution incident in the study region is likely to increase in the future, especially if the region becomes a major transhipment center for crude and refined oil. The extent of this threat will be rooted in the amount of trade within the region, the potential increase in imports and exports, and the number of offshore drillings. Of these three sources, offshore drilling probably poses no threat so long as current policies restricting exploration and drilling are enforced. Should these policies be reversed, exploration could continue; however, oil fields of considerable magnitude would have to be found before drilling would be economically feasible. In the interim it is possible that better safety and containment devices will be developed and the potential threat reduced. It is also possible, however, that as other uses of salt water estuaries, (such as recreation) increase, the possible damages from a blowout would be greatly magnified.

Increase in Tanker Traffic

A more immediate threat is the rise in the number of ships expected to frequent waters in the region and the amount of refined products likely to be transported in the inland waters of Puget Sound and the Strait of Georgia. As population and industry grows within the region so will the demand for refined oil. This oil will have to be transported, either by pipeline or by water. The petroleum product demand in British Columbia is expected to increase from 39,118 million barrels (m/b) in 1966, to 53,940 m/b in 1975, to 65,470 m/b in 1980 to 80,160 m/b in 1985 and to 92,890 m/b in 1990.[14] Much of this demand will be consumed on the mainland; however, unless refineries and/or pipelines are built to service Vancouver Island and upcoast stations there will be a proportionate increase in the amount of oil transported by water. In Washington State the average compound rate growth trend for petroleum is expected to average between 3.0 percent and 3.2 percent annually for the period 1960 to 1985, and corresponds fairly closely to the percentage increase in British Columbia.[15] The internal waterborne movements of refined petroleum, however, is not expected to increase proportionately. In 1952, 2,300,000 tons of refined petroleum were transported in Puget Sound. The tonnage rose to 9,800,000 tons in 1963 and then decreased to 6,300,000 tons in 1966. The trend prior to 1963 indicated an average annual rate of growth of 475,800 tons annually yielding 16,000,000 tons in 1980, 25,600,000 tons in 2000 and 35,100,000 tons in 2020. This trend was reversed by the construction of the Olympic Pipe Line and one study forecasts that future tonnage of waterborne refined petroleum will average 6,300,000 tons yearly until 2020.[16] Such a projection makes an implicit assumption that pipelines will be constructed to move the increased production from refineries within the state. Given that internal transport of refined products approximates existing tonnages, the oil spill potential still increases as more and more vessels use the Sound and increase the possibilities of collision. A study by Honeywell predicted from 2 to 4 marine collisions in the next ten years in Puget Sound.[17] While alarming, these figures did not take into account marine traffic destined for Victoria and some local ship and barge movements in Washington and British Columbia. The prediction may, therefore, be conservative, and the potential of collision will be even greater if the pipelines are not constructed and more refined products move by barge and tanker to markets.

The Problem Posed by Alaskan Oil

The total amount of crude oil entering and leaving the region is also expected to increase, with the amount depending largely on the amount of Alaska crude entering Puget Sound. Estimates of oil reserves in the Alaska fields vary from a conservative estimate of 5-10 billion barrels

up to 30-40 billion barrels.[18] Various methods have been suggested to
move Alaska crude. These include: (1) tankers to move oil through the
Arctic to markets in Europe and the continental United States, (2) a pipeline through Canada to serve markets in the American midwest and (3) a
pipeline from the Alaskan north slope to Valdez and then by tanker to
Pacific Rim markets such as Japan and the west coast of the United States.
Of these the Valdez route has been the most widely discussed alternative.[19]
A Report to the Federal Water Pollution Control Administration quoted estimates of flows that varied from 500,000 barrels per day in 1972 to 2.5
million barrels per day by 1980 as a low estimate and 1 million barrels
per day in 1972 to 5 million barrels per day in 1980 as a high estimate.[20]
The Valdez route will not be completed by 1973; however, if we accept
the 1972 estimates as approximates of the initial flow through the pipeline
and make the assumptions that:

> (1) all oil from Alaska will be shipped to Puget Sound by large
>   tankers,
>
> (2) no oil will go to Japan, California, or the eastern United States
>   without first entering Puget Sound,
>
> (3) pipelines and/or smaller ships will be used to transport oil east or
>   to refineries in Oregon and California,
>
> (4) American vessels will transport oil as specified under clauses of
>   the Jones Act,
>
> (5) the size of the vessels transporting oil will be of the order of
>   250,000 tons,
>
> (6) the 1980 estimates approximate flows eight years after oil ini-
>   tially passes through the pipeline,

some measure of the potential impact is possible. Under the low estimates,
two tankers per week would visit Puget Sound upon initial flow, rising to
twenty tankers per week eight years later. Under the high estimates, four
tankers per week would visit Puget Sound upon initial flow rising to forty

tankers per week eight years later. In addition there would be a number of smaller tankers carrying crude or refined products to California, the number depending on the amount of local consumption and volume of oil transported by pipeline. Puget Sound is one of the few areas on the west coast of the United States capable of handling 250,000 ton tankers. It is at least a day's run closer to Alaska than ports in California. Facilities to handle large ships are limited however and most relevant transportation connections are located on the east side of the Sound.

Atlantic Richfield is the only company that has announced firm plans to utilize oil from Alaska in Puget Sound. Three single bottomed 120,000 ton tankers have been ordered to supply their Cherry Point Refinery with crude oil. Incoming ships would enter the Strait of Juan de Fuca, proceed through American waters in Rosario Strait and then to the Cherry Point Refinery. Outbound ships would leave either through Canadian waters in Haro Strait or back through Rosario Strait before proceeding out through the Strait of Juan de Fuca. These plans are tentative but they do suggest that preliminary estimates of the number of ships likely to enter the Sound may be small, since ships on order are smaller than those projected in earlier estimates. The planned routes for the tankers hug the international border and a spill from just one 120,000 ton tanker could dwarf any international damage that might result from a local tanker or barge.[21]

The potential for greater trade in crude oil and refined petroleum in Puget Sound and the Strait of Georgia should not permit us to underrate the threat of oil pollution that now exists. Existing shipments by barges and tankers are capable of severely polluting waters of the study area. If oil from Alaska is brought into the Sound it will only intensify the potential for spillage, not create a new threat.

The Perimeters for Examining Domestic Legislation and International Treaties

The study region as defined in Chapter One lies entirely within the territorial waters of the United States and Canada. Under international law both nations may exercise almost complete jurisdiction over their territorial seas, if not include these waters within their own national boundaries. They may also exercise control over all activities in the territorial seas save for a few exceptions such as the right of innocent passage.[22] Domestic laws can be utilized to prevent and control actual and potential threats from oil spillage in each nation's territorial sea.[23] International conventions on oil spillage apply insofar as they are incorporated directly into domestic law or are ratified by Canada and the United States (see Table 8 for an outline of major developments to prevent oil pollution at the domestic and international levels).

The waters of Puget Sound and the Strait of Georgia are under domestic jurisdiction; however, their spatial position is not static. The water is domestic in the sense that each sovereign entity can exercise control over its use. However, it exhibits all the characteristics of an international common property resource because of water circulation within the region. Water circulation in Puget Sound, the Strait of Georgia, and on the high seas, can result in international problems if oil spilled in one jurisdiction spreads to other jurisdictions, thereby creating an incident which is beyond the scope of existing domestic law or international treaties.[24]

International Problems in Boundary Areas

When international problems have arisen along the boundary between Canada and the United States in the Pacific Northwest, it has usually been possible to placate differences and resolve the issue given three conditions. First, technology has been able either to solve the problem or harness the resource. Second, unilateral domestic action could offset the problem. Third, both nations recognized that a problem existed and some international mechanism was created to resolve the differences.[25] Technology does not exist for controlling water circulation within the

TABLE 8

OIL POLLUTION-LAWS, CONFERENCES, AND ORGANIZATIONS -
A SUMMARY

Year	International	United States	Canada
1868			Fisheries Act introduced. Amended in 1886, 1894, 1895, 1906, 1914, 1927, 1932, 1952, 1960-61, 1968-69, 1969-70.
1886		New York Harbor Act	Navigable Waters Protection Act
1897	Comité Maritime International (CMI)		
1899		Refuse Act	
1917			Migratory Birds Convention Act
1924		Oil Pollution Act (Original)	
1926	Conference on Oil Pollution of Navigable Waters, Washington, D.C. (Never Ratified)	Organized 1926 Conference	Attended Conference
1948	Convention on the Intergovernmental Maritime Consultative Organization, (IMCO), Geneva	Extension of Admiralty Jurisdiction Act	
1950		Federal Disaster Assistance Act	
1953		Outer Continental Shelf Lands Act	

TABLE 8
cont.

Year	International	United States	Canada
1954	Convention on the Prevention of Pollution of the Sea by Oil, London		
1956	-		Canada accepts 1954 Convention and incorporates it into the Canada Shipping Act
1958	United Nations Conference on the Law of the Sea, Geneva. Convention on the Territorial Sea and the Contiguous Zone. Convention on the Continental Shelf. Convention on the High Seas. The 1954 Convention comes into force.		
1959	International Oil Conference, Copenhagen Preparatory work for 1962 Conference		
1961		Oil Pollution Act (Implemented 1954 Convention)	
1962	Second London Conference on Prevention of Pollution of the Seas by Oil (amended the 1954 Convention)		

TABLE 8
cont.

Year	International	United States	Canada
1965			Canada accepts 1962 Convention and incorporates it into the Canada Shipping Act
1966		Oil Pollution Act (implemented 1962 Convention) Clean Waters Restoration Act (amended 1924 Act)	
1967			Pollution Control Act, British Columbia
1969	IMCO Draft Conventions. Right of Coastal States to Intervene in Casualities on the High Sea. Civil Liability for Oil Pollution Damage.		
1970		Water Quality Improvement Act, 1970 (Supercedes 1924 Act) Washington Oil Spill Act, 1970	Arctic Waters Pollution Prevention Act Interim Federal Contingency Plan released
1971	International Convention on the Establishment of an International Fund for Oil Pollution Damage		Canada Shipping Act amended

* Both Canada and the United States have National Maritime Law Associations as part of CMI.

Source: Based on N.D. Shutler, "Pollution of the Sea by Oil," Houston Law Review, No. 7, (March, 1970), p. 422.

estuary, unilateral domestic action could not internalize all the costs associated with oil spillage, and there is no mechanism for resolving salt water boundary problems.

Precedents exist under international law for holding one state liable for water pollution caused by another state under certain conditions, but these precedents have been used sparingly because of the reluctance of states to surrender autonomy or part of their fiscal resources to obtain the long range benefits from clean seas and shores.[26] Liability has been limited largely to cases where extraterritorial effects have caused injury to beneficial uses. The Trail Smelter case, which involved air pollution damage in the United States from the Trail Smelter in Canada, has been the leading precedent for state to state liability for injurious water pollution. It has also been interpreted as imposing a duty upon nations to prevent damage to the beneficial uses of water in neighbouring states.[27] State responsibility for extraterritorial damage has been based upon the concepts of neighbourship, abuse of rights, and international servitudes as they have evolved in cases of international river pollution.[28] Most of these concepts, while explaining why agreements may have been reached, offer no prescriptive solutions to pollution problems in estuarine or ocean areas.

In the Trail Smelter case, however, the Tribunal regarded the analogy between air and river pollution as close and cited precedents concerning river pollution law in deciding that the Trail Smelter was liable for damage suffered in the United States. The analogy is even closer between air and ocean pollution because of their international common property characteristics. The Trail Smelter case warrants close scrutiny as a model for state liability for extraterritorial damage.

The Trail Smelter Case

Damages from extraterritorial air pollution were beyond the power of domestic jurisdictions. Washington residents affected by the smelter emissions could not bring suits in British Columbia courts because it was beyond the power of the courts to rule on damage to land outside the province and Washington courts could not invoke penalties against a foreign corporation. Smelters usually acquired smoke easements from owners of lands affected or threatened by fumes but this course was not available because the Constitution of the State of Washington provided that no alien person or corporation could hold interests in land in the State. These difficulties were surmounted by transferring individual claims against the Trail Smelter into an international tort between states.[29] The Tribunal held, as a matter of general international law, that extraterritorial damage from pollution is cause for action only between sovereign states and that a state has responsibility for extraterritorial injury.

Despite objections from the United States, the Trail Smelter Tribunal refused to consider damages for violation of the sovereignty of the United States although it did award damages for injury to specific properties.[30] Other international conflicts such as the Lake Lanoux and Corfu Channel cases have also dealt with infringements of sovereignty. In both instances the decisions confirmed that while a state has responsibility for extraterritorial damage, this responsibility does not imply absolute liability for all damages or for infringements of sovereignty.[31]

International Tribunals

While international law recognizes the responsibility of states for extraterritorial damages, international judicial processes do not provide adequate means for obtaining compensation for damages and securing "the discontinuance of the injurious activity or its prevention even before actual damage is suffered."[32] An international tribunal could act only with the consent of the offending state. Recourse to the International Court of Justice or to some third party settlement procedures would be binding only if both parties agreed to compulsory arbitration.

The remainder of this chapter analyses the respective efforts
of the United States and Canada to prevent and control oil pollution in
their territorial waters and to avoid extraterritorial damage. The respective legislation, regulations, plans, and institutions designed to minimize
damage from oil spills are outlined. This is followed by an evaluation of
these efforts in view of the potential threat from an international oil spill
in the study region. The evaluation focuses on two questions. First, how
closely do existing efforts of the United States and Canada internalize the
technological externality through a system of effluent charges as suggested
in Chapter Two? Second, are existing efforts adequate in view of the present state of international law?

THE UNITED STATES-WASHINGTON RESPONSE

The United States-Washington response to the threat of international oil spills has been characterized by a concern with oil after it has been
spilled. Governments have relied on private remedies through laws relating to trespass, nuisance, and negligence to compensate persons suffering
oil damage. Efforts have been made at the federal level to limit the effect of spilled oil, and specific legislation has been passed prohibiting discharges and establishing liability for persons spilling oil. More recently,
Washington State has taken an active role through programs designed to
prevent and control spilled oil under legislation passed in 1969 and 1970,
and through establishment of the Washington State Department of Ecology.
Few laws have been passed establishing funds for cleanup, or compensating
victims of spilled oil, and those that have have restricted liability clauses.
Even less attention has been given to repercussions which could result from
international oil spills, the excuse being that existing laws and contingency plans are not yet sufficient to provide for domestic needs.

Both federal and state statutes are applicable to oil spills occurring
in or entering the navigable and interstate waters of Washington State.[33]

Administrative and legal difficulties can arise in prosecuting violators of these statutes under a concurrent jurisdiction; however, the federal government's stated policy is to surrender authority to states as they acquire capabilities to prevent and prosecute violations. To reduce the possibility of double prosecution, the Washington State Department of Ecology, the United States Coast Guard, and the Environmental Protection Agency, Northwest Region are coordinating programs "very closely with respect to response, cleanup and enforcement programs, so as not to overlap ... administrative responsibilities."[34] The federal government plans to maintain an active role so that it may respond to spills which are beyond the capabilities of the state to control, either because of their size and areal extent or their possible extraterritorial effects. As noted earlier, the national government retains ultimate responsibility for any extraterritorial pollution emanating from within its sovereign jurisdiction. Presumably this means that an international spill would dictate some response and participation by the federal government; however, at this writing, many of the practical problems relating to extraterritorial matters have not been resolved by the federal, state, and foreign governments.[35]

Private Remedies

Jurisdiction

Most of the federal statutes restrict discharges, establish sanctions and penalties, and set liability standards. Liability is limited largely to government claims. Private compensation for damage incurred by spilled oil is restricted to domestic courts, but the law is not well established. Plaintiffs suffering damage have the alternative of bringing their claims in federal or state courts. Maritime law, a federal common law, is usually regarded as supreme over state law in the case of a maritime tort.[36] Prior to 1948, maritime admiralty law did not specifically recognize a ship to shore tort such as pollution of a private beach from oil discharged by a damaged vessel. The Extension of Admiralty Jurisdiction Act of 1948 corrected this inequity, but liability for damage to shore installations was limited to that caused by vessels. The 1948 amendments did not affect instances where a vessel or a sea based installation caused damage to marine life, and the act has been used to hold persons liable for polluting and damaging territorial waters.[37] State law is thought to be generally applicable to damage caused by offshore drilling in territorial waters; however, at least one author contends that, in no matter what forum the claim is presented, the general maritime law of the United States rather than the settled tort law of the state in which the court sits, primarily governs the case.[38] No matter what jurisdiction is supreme, laws at the federal and state levels usually require the injured party to prove negligence[39] - a factor which is very disadvantageous to the plaintiff since it requires a court case based on a minimum of information as to the primary reason for the spillage.

Negligence, Trespass, Nuisance

Beyond the liability that might be established in admiralty and maritime courts, three common law principles could be employed - negligence, trespass, and nuisance. All have limited application, since they require proof of causation. Negligence is the principle remedy available to private persons, but it requires proof that the defendant deviated from a particular standard of care. Beach front owners, farmers of the seabed, and resort owners have all collected damages under negligence actions.[40] Compensation based on a claim of trespass is more difficult to sustain. Trespass requires proof of intentional action and actual contact with property before liability can be established. Nuisance charges are a more "appropriate claim since they usually relate to invasion of property rights rather than damage to property itself."[41] Nuisance principles have held "that houseowners along navigable waters who had suffered loss of the use

of the beach and the shore because of an oil spill from a barge were entitled to compensation for such annoyance, inconvenience, and discom-
fort."[42]

Oil Pollution Laws - Historical Development

Prior to 1969, federal laws provided the basis for most actions a-
gainst oil pollution. Some of these laws are still applicable, but they focus primarily on prohibition and prosecution. Attention has been given to
compensation for damages, but this attention has been limited largely to
government claims. The federal laws discussed below provided the basis
for United States efforts at prevention and control until passage of the Water Quality Improvement Act of 1970. These laws include the Rivers and
Harbors Act of 1899 (also commonly known as the Refuse Act and hereafter
cited as such), the Oil Pollution Act of 1924, the Oil Pollution Act of
1961, the Federal Water Pollution Control Act of 1948, and the Clean
Waters Restoration Act of 1966. Acts passed prior to 1961 were concerned
primarily with domestic pollution, although there was an implicit recognition in all the acts of the threat of international problems resulting from
oil pollution of the sea. The Acts passed in 1961 and 1966 explicitly recognize the threat of extraterritorial damage, since they implement provisions of the 1954 and 1962 International Conventions.

Refuse Act, 1899

The Refuse Act of 1899 provided that:[43]

> it shall not be lawful to throw, discharge or deposit ...
> any refuse matter of any kind or description whatever,
> other than that flowing from streets and sewers and
> passing therefrom in a liquified state into any navi-
> gable waters of the United States.

The waters include the coastal territorial waters, extending three miles
seaward, and the Act applies equally to vessels and shore installations.
Violators of the Act are subject to both a fine of not less than 500 dollars

or more than 2500 dollars and imprisonment for not less than 30 days or more than a year.[44] As an aid to surveillance, informers are entitled to half the fines although a Massachusetts court has ruled that the informer should provide the necessary evidence for a conviction to collect his portion of the fine.

Although not originally drafted to curb oil pollution the Refuse Act has been widely utilized to prosecute those who discharge oil into navigable waters. The Act has been referred to the Supreme Court which held that refuse of "any kind or description" forbids a narrow and cramped interpretation of the Act and "oil is oil whether usable or not by industry standards, it has the same deleterious effect on waterways. In either case its presence in our rivers and harbors is both a menace to navigation and a pollutant."[45] Lower courts have also held that the Refuse Act imposes a duty on violators to reimburse the United States for the cost of removing spilled oil.[46] Interest has grown in the Refuse Act and while the courts have interpreted the Act broadly, there is presently no answer as to whether there is a private right of action under the Act.[47]

Oil Pollution Act, 1924

The Oil Pollution Act of 1924 was more directly concerned with oil than the Refuse Act; however, its application was more restrictive. The 1924 Act was the first to prohibit the discharge of oil by any method, means, or manner into or upon the coastal navigable waters of the United States from any vessel using oil as fuel or any vessel carrying or having oil on board in excess of that necessary for its lubricating requirements.[48] Persons violating the Act were subject to the same fines and prison terms provided for in the Refuse Act. In addition the Coast Guard was empowered to suspend or revoke the licenses of masters and other officers of offending vessels. The only defense against such penalties was proof that the discharge was an emergency or unavoidable accident.

Clean Waters Restoration Act, 1966

The Oil Pollution Act of 1924 was amended by the Clean Waters Restoration Act of 1966 and the strict liability clauses of the 1924 Act were amended. The Clean Waters Restoration Act of 1966 broadened the 1924 Act to: (1) prohibit discharges upon adjoining shorelines as well as upon the navigable waters of the United States, (2) remove the word "coastal" from "coastal waters" so that all lakes and rivers were covered in addition to the coastal waters, (3) require the person discharging oil from a vessel to remove oil from waters and shorelines, (4) provide the Secretary of the Interior with authority to arrange for the removal of the oil and make the violator liable for removal costs and additional penalties, and (5) impose a fine of not more than 10,000 dollars against the offending vessel; the fine constituting a maritime lien against the vessel which prevented it from clearing any port until the fine was paid.[49] The 1966 Act also narrowed the definition of discharges and thereby restricted enforcement. To be punishable an oil discharge had to be either "grossly negligent" or "wilfully negligent."[50] Liability exemptions were also extended to cases related to: (1) emergencies imperiling life or property, (2) unavoidable accidents, (3) collisions or strandings, and (4) instances where discharges are permitted by regulations established by the United States Government.[51] These clauses would, for example, exempt the Torrey Canyon from any liability. Prior to 1966 the United States Coast Guard used the 1924 Act rather than the 1899 Refuse Act to prosecute sudden non-recurring violations; however, the 1966 Act reduced effectiveness of the 1924 Act.[52] The government had to prove gross negligence or wilful spilling and this is difficult when most spills are from an unknown source.[53] As a result the Refuse Act evoked renewed interest as a means of prosecuting those discharging oil into the sea.

Water Pollution Control Act, 1948

The Federal Water Pollution Control Act was enacted in 1948 to enhance the quality and value of our natural resources and to establish a national policy for the prevention, control, and abatement of water pollution.[54] The Act requires states to establish enforceable water quality standards applicable to interstate and coastal waters. Standards proposed by the state are subject to approval by the Secretary of the Interior. These standards must protect public health and welfare and enhance the quality of the water. If a state fails to establish enforceable standards the Secretary is empowered to adopt such standards for that state.[55] Standards are good for preventing pollution of a continuing nature, but they are not particularly suited to nonrecurring violations such as oil spillage. The Act has been amended several times, and, while its direct application to oil pollution is limited, it did provide the basis for the most recent oil pollution legislation, the Water Quality Improvement Act of 1970.

The 1954 Convention

After passing the Federal Water Pollution Control Act in 1948 Congress did not pass any specific legislation to control oil spillage until 1961 when it implemented the 1954 Convention. The 1954 Convention was primarily concerned with regulation of deliberate discharge of oil wastes in the course of tank cleaning operations at sea. More specifically, discharges of oil or oily mixtures by tankers were prohibited generally within 50 miles of the nearest land and, in the case of the Canadian west coast, 100 miles from land. Exemptions were granted in cases where discharge was necessary to secure safety of ships, to prevent damage to ship or cargo, and to save life at sea. Violators were also exempt when ship damage or unavoidable leakage was the cause of the discharge provided that all precautions were taken to minimize the effect of pollution. Ships were required to install oily ballast water separators and maintain an oil record discharge book. Contracting parties were to provide facilities for handling ballast water. Enforcement was left to the contracting flag state and, if requested to do so by another government, flag states were obligated to conduct investigations and prosecute if necessary under Convention provisions. Penalties for violation of the Convention were to be at least as strong as if the incident had been committed in the territorial waters of the polluted state.[56]

Oil Pollution Act, 1961

The United States advised and consented to the International Convention in 1961 and implemented its provisions in the Oil Pollution Act of 1961.[57] Attached to the instruments of ratification were two important reservations. Article XI of the Convention was interpreted as effectively reserving "to the parties of the convention freedom of legislative action in territorial waters."[58] This meant that offences in United States territorial waters would continue to be punishable under United States laws regardless of the ship's registry. Article VII holds the contracting nations responsible for providing facilities to accept oily discharges. The United States declared, because of its lack of jurisdiction, that, while the federal government would not be obliged to "construct, operate or maintain shore facilities... or to assume any financial obligations to assist in such activities," it would ratify the 1954 Convention.[59] The Oil Pollution Act of 1961 retained the penalties first established in the Refuse Act of 1899 but added penalties related to installation of equipment and maintenance of an oil record book. Violators using improper equipment were subject to fines of 100 dollars. Failure to maintain an oil record book would result in a fine of not less than 500 dollars nor more than 1,000 dollars and any false entries would be subject to the same fines and/or imprisonment for up to six months.[60]

The 1962 Convention

The 1954 Convention was amended in 1962 and the United States implemented these amendments in 1966.[61] The Convention expanded coverage from tankers of over 500 gross registered tons to those over 150 gross registered tons. The requirement for separators in tankers was eliminated provided that other means were developed to keep oil on the ship. In addition to the 1954 Convention exemptions, discharges of residue resulting from purification or clarification of fuel oil or lubricating oil are allowed provided such a discharge is made as far as possible from land. Prohibition discharge zones were redesigned and extended. In addition penalties were to be set at a level as severe as those imposed in the territorial waters and of sufficient severity to discourage any unlawful discharges. The United States increased penalties providing for a fine of from 500 to 2,500 dollars and/or imprisonment for a duration not exceeding one year. The Coast Guard was empowered to suspend or revoke a license issued to the master or other officer of the offending vessel.[62] Like the 1954 Convention, the 1962 Convention has inherent limitations. Unless close surveillance enables detection of spills in close proximity to the ship, it is difficult to implicate the vessel.[63] In addition, penalties and enforcement procedures are awkward; penalties are left to the state of the flag and are not specified.[64] If no settlement is reached, the only recourse is to litigation in some international forum. Finally, the Conventions are not universal, apply only to ships registered by the signatories, and are enforceable only in their territorial waters.

Existing Oil Pollution Laws

In 1970 both the United States federal government and the State of Washington passed laws relating to oil spillage; however, at this writing prosecutions for violation of existing law are still being undertaken at least partially under federal laws passed before 1967. In 1970, Congress enacted the Water Quality Improvement Act of 1970 which amended the Federal Water Pollution Control Act originally passed in 1948. The 1970 Act expressly repealed the Oil Pollution Act of 1924, leaving the Refuse Act of 1899, the laws ratifying the 1954 and 1962 Conventions, and the amended Federal Water Pollution Control Act as the federal framework for preventing and controlling oil pollution. Washington State enacted laws to control oil spillage in 1969 and 1970 and these are now being integrated with the Water Quality Improvement Act of 1970.

Water Quality Improvement Act, 1970

The 1970 Act prohibits the discharge of harmful quantities of oil as determined by the President, from all vessels and onshore and offshore facilities. The President is authorized to grant exemptions and the Act specifically exempts all vessels of the United States government and those of foreign governments not engaged in commerce from the prohibition provisions. In addition discharge is permitted as specified under Article IV of the 1954 Convention as amended. Prohibited areas include the navigable waters of the United States, the territorial sea, the adjoining shorelines and waters of the contiguous zone.[65]

Duties of the President. Under the 1970 Act the President is given powers in addition to the discharge exemptions. He can order to arrange for the removal of spilled oil if the owner does not clean it up. Such action is not limited to cleaning up the spilled oil; whenever an actual or potential discharge poses a substantial threat of a pollution hazard in any of the prohibited areas the President is authorized to take all action necessary to minimize the threat and/or damage. He is also directed to issue regulations establishing methods for the removal of oil, requiring equipment for prevention of discharge of oil, and governing inspection procedures.[66] Failure to comply with the regulations can result in a fine of 500 dollars.

Under Section II of the Act the President is obligated to prepare a national contingency plan which also takes into account local and regional
needs.[67]

Discharge Penalties. When oil is spilled in the prohibited zones, the person in charge of the vessel or facility must immediately notify the appropriate federal agency. Failure to do so makes the violator liable to a fine
of not more than 10,000 dollars and/or imprisonment for one year. If the
oil is knowingly discharged and then reported the discharge is punishable
by a civil penalty assessed by the Secretary of Transportation of not more
than 10,000 dollars, but there are no provisions in the Act for imprisonment for an illegal discharge. The penalty cannot be assessed until the
individual charged has had a hearing. The Secretary is authorized to compromise the penalty and can withhold port clearance for any vessel owned
or operated by persons violating the Act whether or not the vessel involved
was the offending vessel. This Act amended the 1924 Oil Pollution Act
where clearance could be withheld only for the ship committing the violation. By using criminal sanctions only in the event of failure to report
discharges Congress attempted to remove the stigma attached to criminal
penalties and shift the focus to preventing and mitigating pollution rather
than punishing violators.[68]

Financial Responsibility. Congress further enacted provisions to finance
the enforcement of the Act and establish funds to meet liability claims.
A revolving fund of 35 million dollars was created to carry out provisions
of the Act and the fund was to be administered by the Secretary of Transportation.[69] Since April 3, 1971 all vessels of over 300 gross tons including barges which use any American port or place or the navigable waters
of the United States have had to provide evidence of financial responsibility to meet any liability claims. The limits placed on this liability were

100 dollars per gross ton or 14 million dollars whichever is less.[70]

Liability. Liability clauses formed a major part of the Act and they attempt
to codify owner, operator, and third party responsibility. If discharge is
caused by wilful negligence or wilful misconduct and the owner or operator of the vessel has privity or knowledge of the act, the government is
empowered to recover all costs of cleanup. The government may also collect damages if there is no wilful negligence or misconduct unless the owner or operator can prove the discharge was caused solely by an act of God,
an act of war, negligence on the part of the government, or any act or
omission by a third party. Liability is limited, however. Vessels are required to prove financial responsibility to the extent of liability: 100 dollars per gross registered ton or 14 million dollars, whichever is less. Owners of more than one vessel must establish proof only for the largest unit.
The liability for discharges emanating from onshore or offshore facilities is
8 million dollars; however, owners are not required to establish evidence
of financial responsibility. Third party liability is the same as that for owners or operators except his liability is limited to the amount it would have
been had he owned the vessel. These liability provisions exceed the liability exemptions previously in force under the Limited Liability Act of
1851, the Outer Continental Shelf Lands Act, the Oil Pollution Act of
1924, and the Clean Waters Restoration Act of 1966. The 1970 Act is am-
biguous, however, in that it fails to specify who shall be liable if the owner and operator are not the same person. Ambiguity is also present with
regard to wilful discharges, discharges from onshore and offshore facilities,
and to third party liability. Meiklejohn maintains that where the owner
and operator are separate "it would seem the government could proceed
against either or both."[71]

Assessment of the Act. While the Water Quality Improvement Act of 1970

is a significant improvement over previous legislation, it has several in-
adequacies apart from its inability to deal with extraterritorial pollution
as discussed in part four of this chapter. First, instances could arise in
which 14 million dollars would be insufficient to cover costs of cleanup.
Meiklejohn asserts that the 14 million dollar liability ceiling limits liability to 52,000 tons of oil.[72] Even if the assertion were false, difficulties
would arise in allocating costs in excess of the liability limits of the owner. Such costs would have to be internalized by private persons who are
injured, by the government, by some other third party, or by the owner
or operator.

Of these groups the owner or operator is probably in the best position to meet and distribute the costs. Such persons can be held morally
accountable on the basis of the theory of enterprise liability where business presents: (1) an abnormal danger to the community, (2) a danger that
cannot be eliminated, (3) a danger that intentionally exposes private persons to the business risk, and (4) a profit motive for the owner engaged
in the risk producing activity. Governments can be expected to share
some of the liability, because of their income from taxes and oil leases.[73]
They also have power to intensify inspection and enforcement procedures
and are therefore more able to bear the costs of cleanup and to spread the
risk involved than are private individuals or some third parties. They have
an additional moral commitment to compensate innocent private victims
if improper government equipment, inspection, or enforcement was the
cause of the discharge. Under the Act liability is placed upon the owner
or operator of the vessel. The owner of the cargo is under no legal obligation to pay for cleanup costs. If the cargo owner were held liable for
costs beyond the limits of the vessel owner or operator, it would be possible to cover the total costs of cleanup. This proposal would secure financial responsibility for the total amount of oil transported, and would
result in a more equitable distribution of the costs. While governments

may be able to recover total cleanup costs, the Act does nothing to compensate individual victims. The only recourse to those innocent parties are the private remedies discussed earlier in this chapter, remedies which are very disadvantageous to the plaintiff.[74] The strength of the 1970 Act is its emphasis on cleanup following a discharge. Establishment of cargo owner and government liability beyond the limits of owner or operator liability and compensation for innocent victims of discharge would enhance the effectiveness of the Act to minimize discharge and damage and to equitably distribute the risks and costs of drilling and transporting oil.

Regional Responses to the Water Quality Improvement Act, 1970

While federal legislation has had a paramount effect on preventing discharges, Section II of the Water Quality Improvement Act of 1970 expressly permits states and regional and local governments to impose any requirement or liability with respect to the discharge of oil into any waters within the state. Two states, Maine and Washington, have enacted stricter liability laws and other legislatures, including Oregon, Florida, and Massachusetts, have passed or are considering strict oil pollution control laws. All such laws present serious constitutional and international questions in view of possible conflicts with the Water Quality Improvement Act of 1970, the present international obligations of the United States, future claims for any extraterritorial damage from spilled oil, and with the 1969 and 1971 Brussels Oil Pollution Conventions. It is interesting to note in light of the questions posed by this study that both Maine and Washington are coastal states bordering Canada. Large amounts of crude and refined oil are transported in water proximate to each state and there is a distinct threat of extraterritorial damage from oil pollution.[75]

The Washington legislature has recognized the rights of its citizens to pure and abundant waters and in 1969, 1970, and 1971 enacted laws amending broad anti-pollution statutes passed in 1915, 1916, and 1945 to

protect these rights.[76] These laws sought to enhance water quality by regulating oil discharges. In 1969 the legislature passed an oil discharge law
which specifically prohibited intentional and negligent oil discharges and
held violators subject to a civil penalty of up to 20,000 dollars.

Washington Oil Spill Act, 1970

In 1970 the legislature passed the Washington Oil Spill Act. The
Act provides that except where discharge is: (1) authorized by the Water
Pollution Control Commission prior to entry into state water, (2) permitted
under operation of the Act, (3) caused by an act of war or sabotage, or
(4) caused by negligence on part of the United States Government or the
State of Washington, the entry of oil into waters of the state is unlawful.[77]
If a spill does occur, the person causing such action is immediately required
to remove the oil and notify the State Department of Ecology. The Department in turn agreed with federal authorities that the Coast Guard would be
the agency to contact. Should a person be unable to remove the oil he
must take all practicable means to contain and disperse the discharge. The
Director of the State Department of Ecology and his staff are further authorized to arrange for the removal of oil if private individuals do not comply
with state law and have complete control, even in the instance of private
cleanup, over all materials used. Any person who fails to immediately
collect, remove, contain, or disperse oil when under an obligation to do
so is held responsible for the "necessary expenses" incurred by the state in
performing these tasks.[78] The Director informs parties of the expenses and
they become due and payable. The Water Pollution Control Commission
may, upon application from the recipient of the order, reduce or set aside
all or part of the amount due when it appears "just and fair in all circumstances."[79]

Penalties. The Washington Oil Spill Act of 1970, like the Water Quality

Improvement Act of 1970, does not provide for criminal punishment of violators. However, persons who negligently or intentionally discharge oil
are liable for a civil penalty of not more than 20,000 dollars. This penalty is to be assessed by the Director of the Department of Ecology and, u-
pon application by the violator, the Director may compromise the penalty.
No other penalties for unlawful discharge are provided by the Act.[80]

Liability. The 1970 Washington Act differs from the 1970 federal act,
however, in terms of the liability of violators for pollution damage to persons and property. The Washington Act imposes strict liability without regard to fault for damage to all persons or property, public or private, resulting from the unlawful entry of oil into the waters of the state. The liability is imposed on owners or persons having control over the oil.[81]
There is no limit on the amount of liability facing oil companies or shipowners for cleanup expenses or damages, although, as discussed above, the
Water Pollution Control Commission can reduce the amount of reimbursement for expenses by an amount it considers just and fair. The Act was intended to eliminate the difficulty experienced by claimants attempting to
prove that the injurious spillage occurred as a result of the wilfulness or
negligence of the person controlling the oil. Proponents of the legislation
argued that strict liability would be the most efficient solution to compensate private parties for damages caused by oil pollution to persons and property. First, it would provide individuals with the power to take action in
civil courts where there are no liability limitations and would circumvent
difficulties apparent with admiralty law, where liability provisions are
based largely on proof of negligence. Second, by assigning liability a-
gainst the owner or those having control over oil it avoided the difficulty
of proving negligence on the part of tanker owners or operators and onshore and offshore facilities. Conventional tort actions against foreign
built ships, registered in flag of convenience states, with multinational

crews are almost impossible. Third, the legislature reasoned that, with
the size of the world oil industry and the evolving doctrine of enterprise
liability, the industry was more capable of distributing and accounting for
environmental risks than either private persons or the State of Washington.[82]

Assessment of the Act. The Washington Oil Spill Act provides the Department of Ecology with power to solve problems resulting from major spills;
however, it does little to institute prevention programs and improve the individual's property and environmental rights. The Act seeks to provide remedies for oil damage to the water and beauty of the state. The Attorney-
General, on behalf of the people of the State of Washington, may seek
damages for pollution to wildlife, fish, vegetation, and the quality of the
water, and for the cost of restoring such resources injured by the pollutants.
Cunningham comments on the effect of this provision:[83]

> A citizen whose property or person is injured by
> oil pollution has standing to bring a common law or
> maritime action for damages or injunctive relief.
> However, the individual citizen does not have
> standing to bring an action in Washington courts
> for either damages or injunctive relief against a
> private polluter of public waters where the pollution
> or potential pollution does not affect a present
> substantial interest which is substantially different
> from that of the general public. Nor have citizens
> been empowered to secure a writ of mandamus to
> compel the attorney general, whose action is
> discretionary, to bring an action, either criminal
> or civil, against a private polluter in violation
> of the law.

The Act does make it easier to hold persons liable for damages and cleanup
costs provided that the violator is caught and prosecuted to the full extent
of the law. Such an assumption can be highly erroneous. Many of the
sources of oil pollution are unknown, yet cause extensive damage. Violators are encouraged not to notify the Department of Ecology by the strict

liability provisions and in the absence of an intensive surveillance program
are not prosecuted. Moreover, individuals suffering damage must still go
through long and costly court litigation to recover damages.[84]

Coastal Waters Protection Act, 1971

In 1971 the Washington legislature passed the Coastal Waters Protection Act of 1971. This act contains provisions giving the Department of
Ecology broad powers to inspect oil carrying and processing equipment,
control oil removal, establish contingency plans for different areas of the
state, and deal with emergency situations such as traumatic oil spillage.
The Act also created a coastal waters protection fund to be financed
through penalties, fees, and charges collected for oil pollution violations,
and through a one cent per gallon tax on each gasoline tax refund claimed
by persons using petroleum for marine purposes.[85] The State of Maine has
enacted parallel legislation through two laws designed to protect its coast
from oil spills[86] and Michigan has passed a law furthering the environmental rights of its citizens. The first Maine law gives the Environmental Improvement Commission power to license all facilities and vessels involved
in transferring oil within the state and establishes a 4 million dollar Maine
Coastal Protection Fund to meet costs of administration, research, equipment, cleanup of unexplained spills, and third party damages.[87] The fund
is to be created by a one-half cent per barrel tax on all oil coming into the
state. The intent of the 1971 Washington Act is similar to the Maine statute, but it has severe limitations. Claims against the fund can be made for
government expenses only. Unlike the Maine legislation no compensation
is provided for persons who might suffer third party damages or damages
from spills of unknown origins. In addition it is questionable if the method
of funding will be sufficient to meet the costs associated with a major spill.
There is a distinct need in Washington State for a compensation fund for innocent victims of oil spillage and a more adequately financed cleanup fund.

The second Maine law deals with site locations of developments substantially affecting the environment. The Act declares that it is the intention of the legislature to provide for the economic and social well-being of its citizens by controlling the location of commercial and industrial development in order to protect the natural environment of the state. The state is given power to regulate site locations for large edifices such as oil refineries.[88] Washington State now lacks legislative authority to regulate siting of industry on a statewide basis, although initiatives and legislation have been advanced which would give the state some authority in coastal planning. However, neither the initiatives nor existing law gives Washington power equal to that exercised by state authorities in Maine.

The Michigan law creates standing for any legal entity to initiate an action for relief against any other legal entity "for the protection of the air, water and other natural resources and the public trust therein from pollution, impairment or destruction."[89] In addition the Act gives the courts power to review anti-pollution standards to determine their validity, applicability, and reasonableness, and, if evidence warrants, to order adoption of court approved or specified standards.[90] If this Act were adopted in Washington, if plaintiffs were given the right to subpoena information on oil discharges, and if the burden of proof were shifted to the polluter, the environmental rights of Washington citizens would be immeasurably improved.[91]

Contingency Plans

The Water Quality Improvement Act of 1970 instructed the President to prepare and publish a National Contingency Plan within 60 days after signing the bill into law on April 3, 1970. The plan was to include, among other things: (1) establishment or designation of strike forces, (2) a system of surveillance and notice, and (3) provision for regional and state plans. On July 20, 1970 the President issued Executive Order 11548 directing the Secretary of Transportation to assume responsibility for the National Contingency Plan. The Secretary in turn designated the Coast Guard as the operational unit to implement the national plan and develop parallel regional plans. Accordingly, on December 1, 1970 the Commander of the Thirteenth Coast Guard District issued the plan for the Seattle Coastal Region.[92] This regional plan, and, more specifically, the Sub-Regional Plan for Washington is the applicable plan for all United States waters in the study area.

Regional Contingency Plan

The Regional Contingency Plan establishes procedures for reporting oil spills and notifying the Coast Guard. It also provides for the identification, procurement, maintenance, and storage of equipment and supplies. The Plan calls for the establishment of a Regional Response Center and Team in Seattle, which is to be closely linked to and integrated with the National Response Center. Strike and emergency task forces are established to combat particular spills and the plan includes procedures and techniques to be employed in dispersing oil. A schedule identifying dispersants and other chemicals that may be used in carrying out the plan, the waters in which they may be used, and quantities which may be safely applied is provided to aid task forces combatting spilled oil.

A primary focus of the regional plan is provision for a federal response capability at the regional level should individuals and/or states be unable to respond to a particular spill. The plan also seeks to coordinate the responsibilities and facilities of other federal agencies and departments when a federal response is required. It expressly provides for state and local participation in the plans and makes an initial attempt to gather and list information as to equipment and resource capabilities at the state and local level.

The Captain of the Port, Seattle, Washington, is the on-scene commander for the Washington Coastal Sub-Regional Response Center, the area under investigation in this study. All reports of pollution or potential pollution are directed through his office or the nearest Coast Guard Unit, and all communications during pollution incidents are over circuits of the United States Coast Guard. It is the duty of the on-scene commander to inform the authorities designated in the plan and the Washington State Department of Ecology of the spill, and to initiate enforcement procedures.

The Seattle Coastal Regional Contingency Plan authorizes the use of Coast Guard personnel, facilities, and equipment during a pollution incident commensurate with the threat to life and property. It also outlines procedures for utilizing equipment and personnel from other federal departments and agencies. In addition, the plan lists private individuals and companies who have equipment which could be employed to combat oil spills.

Assessment of the Contingency Plan

The federal plans are carefully spelled out on paper; however, the plan is not at full operational capability. Officers in charge of oil pollution lack extensive training in combatting spills. The Coast Guard has had to rely on visual surveillance methods and will not be able to move to electronic sensors until fiscal year 1974. It now maintains aircraft for surveillance, but hopes to replace these planes with more mobile helicopters in fiscal year 1972. Helicopters would be used for surveillance, but could also be of assistance in the deployment of cleanup materials and personnel, should a spill occur. The Coast Guard is currently increasing its response capability by acquiring equipment and materials to respond to a particularly massive oil spill; however, present technology discourages heavy investment in equipment because it does not provide for the complete recovery of spilled oil. First priority is being given to cleanup materials for in-shore spills, and to the development of cleanup devices and systems. To this end, the Coast Guard has established strike forces for the West Coast and is developing plans to place smaller task forces in major ports. If the development program is successful, the Coast Guard hopes to acquire containment and recovery systems for offshore coastal waters and the high seas, but does not expect the systems to be fully operational until 1975.[93]

State and Local Input. Operational capability is further limited by a lack of input from state and local authorities and industry. The Coast Guard is relying on these interests to acquire materials and equipment to combat most minor spills. Washington State did not publish its oil spill action plan until January, 1972.[94] The plan is more detailed than the federal plan in that it is more specific on critical water areas, the types of dispersants which may be used, and the methods of deployment. It also provides a more comprehensive list of available material and equipment which could be used to combat spilled oil. The list is by no means complete, since local and port authorities have yet to develop strike forces which are to form the backbone for the state and regional contingency plans. Many local authorities interpreted the Water Quality Improvement Act of 1970 and subsequent contingency plans as placing responsibility on the Coast Guard to clean up any oil spills and these authorities have made no effort to develop local plans until requested by the Washington State Department of Ecology and the Coast Guard. Industry has been reluctant to invest large amounts of money in containment and recovery systems because of the rapid changes in oil cleanup technology.[95] They are also unsure of the magnitude and type of systems that may be needed in the absence of any firm projections of oil deliveries to Puget Sound from Alaska.

Prevention of Spills. Contingency plans, by their very nature, are responses to pollution problems that have occurred and are threatening life and property. The plans have given little consideration to preventing oil pollution, although the President has proposed legislation which could reduce this threat. The Port and Waterways Safety Bill would provide the Coast Guard with permanent broad authority to issue regulations necessary for the control of ship movements and operations as well as for the supervision of cargo movements at the ship-shore interface. Additional funding has been requested for an Harbor Advisory Radar System, which would aid in ship traffic control in congested waterways; however, it is not fully operational. In the spring of 1971 Congress voted 1 million dollars to begin work on the project in the Puget Sound region. First priority is being given to a radio communication network and later appropriations will support installation of surveillance radar and other sophisticated electronic equipment.[96]

The Contingency Plan and Extraterritorial Pollution. Neither the contingency plans nor the proposed legislation address themselves to the threat of extraterritorial pollution. Contacts have been made between the Coast Guard and the Department of Transport and the National Harbours Board in Canada over specific oil spills, but no formal meetings have been held to coordinate contingency plans or to regulate ship movements in Puget Sound and the Strait of Georgia. In part this negligence stems from the absence of a massive oil spill near the boundary which would be beyond the cleanup capacity of the subnational and local governments. It is also rooted in an emphasis on containment and recovery to the neglect of prevention efforts, in the newness of the legislative mandate, and in the stated necessity of codifying and developing domestic needs and capabilities before incurring international obligations. Section three of this chapter analyzes the Canada-British Columbia response to the threat of extraterritorial damage from oil pollution in the study area within the context of what has been largely a domestic response from the United States and

Washington.

THE CANADA-BRITISH COLUMBIA RESPONSE

The Canada-British Columbia response to the threat of extraterritorial pollution from spilled oil has been precipitated largely by three major events in 1969 and 1970: the voyages of the Manhattan from the East-
ern United States through the Arctic Archipelago to Prudhoe Bay and return; the sinking of the Arrow off the coast of Nova Scotia; and the proposed shipment of Alaska crude by tanker off the British Columbia coast
from Valdez to terminal facilities in Puget Sound. Prior to these events,
Canada lacked specific oil pollution legislation at both the federal and
provincial levels, save for the laws implementing international obligations
such as the 1954 and 1962 International Conventions. Legislation has been
passed by both governments which has a bearing on oil pollution of the
sea, but this legislation concerns specific matters over which the federal
and provincial governments have jurisdiction, and oil pollution controls
have only been treated as incidental features of the more specific legislation. These laws focus on penalties for discharge of oil and make provisions to secure pollution permits but make no reference to compensating
victims or assessing cleanup costs. As in the United States, reliance has
been placed on the common law relating to trespass, nuisance, and negligence to compensate victims incurring damage from spilled oil. During
and after the events of 1969 and 1970, the Canadian federal government
passed the Arctic Waters Pollution Prevention Act, amended the Canada
Shipping Act, and announced efforts to develop Regional Task Forces and
a National Contingency Plan to control oil pollution. These acts and the
regulations promulgated therein constitute the Canada-British Columbia
response to the threat of international pollution, as British Columbia plans
no change in its response until the federal programs have been implemented.
These laws provide greater equity for individuals than most laws in the

United States through establishment of a cleanup fund and compensatory procedures for selected persons; however, many citizens, especially those suffering indirect damages, are not directly compensated under the law, and application of the acts is therefore limited. Contingency plans, on the other hand, are more advanced in the United States than in Canada, where existing plans are still paper creations. Concern over the threat of extraterritorial pollution has been greater in Canada than in the United States since plans were announced to move oil down the British Columbia coast; however, few practical steps have been taken to reduce the threat. Canadian plans are still in the developmental stage. The Cabinet has yet to announce any firm policy regarding oil shipments by tanker along the Pacific coast. Many Americans see little threat of extraterritorial pollution from Canada. Some are unwilling to curb development which might help depressed economies in the states of Washington and Alaska, and the United States maintains that existing laws and contingency plans are not yet sufficient to provide for domestic needs. In this vacuum, the potential for major extraterritorial oil spill damage in the region continues to increase.

Jurisdiction Over Oil Spillage and Pollution

Jurisdiction over oil spillage in Canada is complicated by the British North America Act. Sections 91 and 92 of the Act outline the jurisdictional responsibilities of the federal and provincial governments. The federal government under Section 91 has responsibility to make laws for the peace, order, and good government of Canada; power over trade, commerce, navigation, shipping, criminal law, seacoast and inland fisheries; and control over connecting works and undertakings such as interprovincial railways, canals, and pipelines. Similarily, under Section 92, the provincial governments have control of public lands of the province and responsibility for local works and undertakings, property and civil rights in

the province, and all matters of a local or private nature within the province.[97] Under Section 91, the residual power was left to the federal
government. In part this power allocation was based upon the unitary principle of federalism which recognized that all governments drew their authority from the Crown and differed from the American concept of federalism
where two sovereign entities divided between themselves the various fields
of political activity.[98] The British North America Act is unclear, however,
over the precise nature of the federal union - that is, should the union
stress strong central powers or the rights of the provincial governments?
The validity of each view has often been tested in the Courts. Until 1949,
the final court of appeal was the Judicial Committee of the British Privy
Council. Over the years, the Privy Council demonstrated a marked bias
in favor of provincial rights at the expense of federal power. In 1949, the
Supreme Court of Canada became the final appellate court for jurisdictional disputes. Recent decisions of the court, such as the Reference Re: Offshore Mineral Rights of British Columbia,[99] have tended to be less rigid,
giving the federal government greater jurisdictional responsibilities over
natural resources. The court has been reluctant to rule that any level of
government does not have jurisdiction over a specific problem or area,
leaving it to the politicians to solve their own problems. Even with this
relaxation, the earlier bias remains; a bias that has been particularly important to natural resource development and pollution control, since it has
placed limitations on the ability of the federal government to undertake
projects where it lacks a concurrent responsibility. Often federal-
provincial agreements are necessary to effectively control pollution where
there is a divided responsibility; however, issues of environmental pollution are not debated on their merits. Environmental problems constitute
only one area of federal-provincial disagreement. While the federal government may have some constitutional right to take action, it may not
choose to do so because such action might disturb delicate negotiations

with a province over such matters as constitutional reform, regional economic development, and inflation.[100] As a result the federal government
has been reluctant to intervene until pollution has reached a crisis and the
national interest demands action. Restrictions in the constitution and difficulties in concluding federal-provincial agreements tend to foster post
facto control of pollution.[101]

Extraterritorial Oil Pollution

Oil spilled on the high seas or in the territorial waters of the United States which damaged Canadian waters would require action by the
federal government because it is the only qualified agency recognized in
international law. Any claims for damages would therefore require the
concurrence of the federal government. On the strength of the Reference
Re: Offshore Mineral Rights of British Columbia, it would seem that the
federal government would have jurisdiction seaward of the "ordinary low
water mark" and outside of "harbours, bays, estuaries and other similar inland waters." However, once the oil was deposited above the "ordinary
low water mark" it would seem to fall under provincial jurisdiction and
thus any claim for damage would require consultation with the provincial
government. The jurisdictional problem is further complicated by the issue
of "inland waters," as the terms of reference were limited to a determina-
tion of the jurisdiction seaward of the ordinary low water mark and outside
of harbours, bays, estuaries, and other similar inland waters.[102] To protect its drilling rights (as discussed earlier in this chapter) the provincial
government contends that the Strait of Georgia qualifies as "inland water"
and so comes under provincial jurisdiction. A second position holds that
"under the Canadian Constitution, the matters of navigation and offshore
waters are under federal jurisdiction" and that this permits the federal government to prepare contingency plans and deal with oil spills once they
occur in waters such as the Strait of Georgia. The provincial government

would retain the right to initiate supplementary measures if the federal plans proved inadequate.[103] Thus the province seems willing to surrender the costs of controlling oil pollution in inland waters to the federal government but remains insistent that royalties from offshore drilling should accrue to the provincial treasury. The jurisdictional position of "inland waters" is open to negotiation. In the absence of a federal-provincial agreement to reduce jurisdictional confusion, no conclusive case can be made for the exclusive responsibility for and control of oil pollution by either level of government.

Private Remedies

Because of this jurisdictional dispute, the relatively recent threat of major oil pollution on Canadian shores, and the absence of specific statute law to control oil pollution, common law has had greater significance for public and private bodies in Canada than in the United States. Canadians have relied heavily on common law to collect costs expended in combatting oil spills and to compensate those suffering direct and indirect damage from spilled oil. Common law in Canada, however, is just as unconducive to collecting for damages as common law in the United States. Canadian courts, relying on British precedent, usually have required plaintiffs to prove negligence before costs and damages can be awarded, and have been reluctant to extend compensation to persons suffering damage on the grounds of remoteness.

Admiralty Law

Admiralty law in Canada regarding oil pollution claims is not well established, but Canadian Courts of Admiralty exercise largely the same jurisdiction as the English Courts of Admiralty. Plaintiffs are usually required to prove negligence on the part of the polluter. In one case, however, the Admiralty Court indicated that proof of an unseaworthy condition might also be used to collect damages.[104] A shorefront owner (Southport Corporation) had suffered damage from oil spilled by a tanker, and the shorefront owner claimed negligence on the part of the master and trespass, nuisance, and damage resulting from such negligence. The master was acquitted on negligence, thus the whole claim failed. In its decision, the Court intimated that a different result might have been possible had the shorefront owner claimed the vessel was unseaworthy. This has been the only case of any significance regarding oil pollution claims. However, no suits based solely on a condition of unseaworthiness have ever been successful. Compared to the United States (where several suits have been successful and where persons suffering direct and indirect damage have been compensated), Canadian admiralty law offers little precedent for securing compensation.

Tort Actions

Beyond claims in admiralty, there have been a small number of tort actions by persons under the headings of nuisance, negligence, and trespass, but decisions in these cases cast doubt on the ability of common law to provide adequate compensation. In the decision on ESSO Petroleum Company Ltd. vs. Southport Corporation, for example, the presiding judge considered trespass and nuisance were good causes for action but upheld defence contentions that the plaintiff was in the same position as a person whose land adjoined a public highway and was not entitled to damages unless it could be shown that the damage was done negligently. This decision has cast doubt on the effectiveness of trespass as a means of redress in oil pollution actions. Nuisance would seem to suffer a similar fate if it were not for a Supreme Court decision in City of Portage La Prairie vs. B.C. Pea Growers Ltd.[105] The Court seemed to suggest that negligence is not required to uphold a nuisance claim. While the law regarding nuisance is very unsettled, a claim based on nuisance "may yet be a viable remedy for an indirect non-negligent interference with the use and enjoyment of land."[106] Under such restrictions, it is doubtful whether any person would be held strictly liable for an oil pollution incident.

Actions based upon negligence, the strongest form of common law action, must meet demanding standards. The plaintiff must show that the polluter deviated from a particular duty or obligation recognized by laws which require the polluter to conform to a certain standard of conduct for the protection of others against reasonable risk. Plaintiffs must also demonstrate a reasonably close connection between the conduct and the resulting injury, and must show actual loss or damage resulting from the conduct of the polluter. As a result, indirect damages have seldom been awarded in cases based on negligence.[107] In the United States beachfront owners, farmers of the seabed, and resort owners have all collected oil pollution damages based on suits of negligence. Because Canadian courts have defined a "reasonably close connection" narrowly there have been no successful claims for indirect oil pollution damages in Canada.

Common law remedies of nuisance, trespass, and negligence have not adequately protected interests of oil pollution claimants in Canada. As a result persons suffering damages from oil pollution have been forced to bear the entire costs, while those responsible have no liability unless they acted negligently. Such a situation is neither equitable nor fair. Many lawyers argue that it would not be appropriate to alter the common law to meet the special needs of one type of claimant. The only alternative, therefore, is through statute law, where legislation would protect all persons having an interest in the marine environment and adjacent lands, and provide for equitable means of compensation.

Relevant Federal Laws for Oil Pollution Abatement and Control

Prior to 1971, Canada lacked specific oil pollution control laws. Federal legislation had been passed regarding shipping, navigable waters, fisheries, migratory birds, national harbours, and the criminal code, all items over which the federal authority has a complete or a concurrent jurisdiction. Each of these pieces of legislation has sections which can be applied to oil pollution control, but none was passed primarily to prevent oil pollution damage. Like many of the early United States laws on oil pollution, these acts focus on prohibition and prosecution. International conventions regarding oil pollution have been ratified and incorporated into domestic law; however, Canadian concern for extraterritorial damage from spilled oil is largely a feature of the 1970's. The federal laws discussed below provide the basis for Canadian action prior to the 1971 amendments to the Canada Shipping Act. These include the Canada Shipping Act, the Navigable Waters Protection Act, the Fisheries Act, and the Migratory Birds Convention Act. British Columbia legislation is examined separately. However, as with early federal laws, controls over oil spillage are just one item in legislation primarily designed to serve other needs. These laws include the Health Act, the Water Act, the Pollution Control Act of 1960 and its successor the Pollution Control Act of 1967. British Columbia currently lacks specific oil pollution legislation such as the Washington Oil Spill Act of 1970 and the Coastal Waters Protection Act of 1971.

Fisheries Act

One of the first laws enacted by the Canadian Parliament after Confederation was the Fisheries Act of 1868. Section 14 prohibited the discharge of ballast, coal ashes, stone, lime, chemical substances, and other deleterious substances into any waters where fishing is carried on. Persons committing an offence under the section were subject to a fine not exceeding one hundred dollars or imprisonment for not more than two years.[108] Oil is not specifically mentioned as being prohibited, but presumably it could be labelled a "deleterious substance" and discharges would then be subject to penalties. Amendments in 1886, 1894, 1895,

1906, 1914, 1927, 1932, and 1952[109] altered the description of prohibited substances, changed the penalties, and gave the Governor-in-Council
power to prevent or remedy the obstruction and pollution of streams, but
they were no more specific about oil than the original act. In 1961 Parliament gave the Governor-in-Council specific authority to deem any substance deleterious and consolidated the penalties for violation of the Act.
The first offence was subject to a fine of not less than one hundred dollars
and not more than one thousand dollars and/or imprisonment for not less
than one month and not more than six months. Subsequent offences were
subject to a fine of not less than three hundred dollars and not more than
two thousand dollars and/or imprisonment for a term of not less than two
months and not more than twelve months.[110] In 1970, the Act was further
amended and persons were prohibited from depositing or permitting "the
deposit of a deleterious substance of any type in water frequented by fish
or in any place under any condition where such deleterious substance or
any other deleterious substances may enter any such water."[111] The fine
was increased to a sum not exceeding five thousand dollars, but the imprisonment penalties were dropped. The 1970 amendments further provided
that, in prosecuting for an offence, it is sufficient proof of the offence to
establish that it was committed by an employee or agent unless the accused
establishes the offence was committed without his knowledge or consent
and that he exercised all due diligence to prevent the discharge. The
Minister is given power to take action to repair or remedy the condition or
reduce or mitigate any damage to life and property.[112] Finally, while
the Act gives the Governor-in-Council power to specify deleterious substances, oil is not yet listed as such a substance. Moreover the Act is not
designed to control or prevent pollution—it is designed to protect fish and
the fishing industry. However, since prosecutions have been made under
the Act over pulp mill discharges, it seems reasonable that, should oil be
added to the list, similar prosecutions could be made for oil spills,

provided they adversely affect waters frequented by fish.

Migratory Birds Convention Act

Legislation concerning other forms of wildlife have also been used for pollution control, albeit on a very limited scale. In 1916 the United States and the United Kingdom (acting on Canada's behalf) signed a convention for the protection of migratory birds. In 1917 Canada implemented the provisions of the convention through the Migratory Birds Convention Act. Under the legislation, the Governor-in-Council was given power to make regulations "as are deemed expedient to protect the migratory game, migratory insectivorous and migratory nongame birds that inhabit Canada during the whole or any part of the year."[113] Persons violating provisions of the Act or any regulation were originally liable, upon summary conviction, to a fine of not more than one hundred dollars and not less than ten dollars and/or imprisonment for a term not exceeding six months. In 1921 the maximum fine was increased to three hundred dollars.[114] Initially, the regulations issued under the Act contained no special provisions for oil, but since 1948 the regulations have provided that:[115]

> No person shall knowingly place, allow to be placed or in any manner permit the flow or entrance of oil, oil wastes or substances harmful to migratory waterfowl into or upon waters frequented by migratory waterfowl or waters flowing into such waters or the ice covering either of such waters.

The Canadian Wildlife Service has responsibility for the control of water pollution that may adversely affect migratory birds.[116] The Migratory Birds Convention Act cannot be regarded as a major weapon to control extraterritorial damage: few prosecutions have been made under the Act for spilling oil; the penalties are small and provide little deterrent; and prison terms are seldom invoked. The Convention is, however, one of the few treaties in North America which recognizes that common property resources do not respect political boundaries.

Navigable Waters Protection Act

The Navigable Waters Protection Act has been used to control water pollution but its application to oil spillage is rather tenuous. Sections
18 and 19 of the Act prescribe that "no person shall throw or deposit, suffer or permit to be thrown or deposited any sawdust, edgings, slabs, bark
or like rubbish of any description whatsoever that is liable to interfere with
navigation in any water, any part of which is navigable or that flows into
any navigable waters," and "no person shall throw or deposit or cause,
suffer or permit to be thrown or deposited any stone, gravel, earth, cin-
ders, ashes or other material or rubbish that is liable to sink to the bottom
in any water, any part of which is navigable or that flows into any navigable water where there are not at least twenty fathoms of water at all
times."[117] Earlier acts and amendments to the Navigable Waters Protection Act prescribed separate penalties for both sections,[118] however, in
1969 these were consolidated, making persons contravening each section
liable on summary conviction to a fine not exceeding five thousand dollars[119] - a penalty similar to that provided in the Fisheries Act. The Act
has not been used extensively to combat oil pollution. The Courts have
not evaluated the meaning of "rubbish." If they interpreted it as broadly
as "refuse" was interpreted under the 1899 United States Refuse Act, the
Canadian legislation might have greater relevance.[120] It would also seem,
however, that the pollution sections are somewhat redundant in view of
similar restrictive penalties available in the Fisheries Act, even though
oil pollution is not specifically prohibited under that Act. It is possible
(but highly improbable) that oil adversely affecting navigable waters
would not damage fish frequenting the same waters.

Canada Water Act

The Canada Water Act suffers many of the same restricted applications. The Act makes no direct reference to the problem of oil pollution,

but Section 8 provides that:[121]

> Except in quantities and under conditions prescribed
> with respect to waste disposal in the water quality
> management areas in question ... no person shall
> deposit or permit the deposit of waste of any type
> in any water comprising a water quality management
> area or in any place under any condition when such
> waste or any other waste that results from the
> deposit of such waste may enter any such waters.

where "wastes" are defined as:[122]

> Any substance that is added to the water, would
> degrade or alter or form part of a process of
> degradation or alteration of the quality of these
> waters to an extent, that is detrimental to their
> use by man or by any animal, fish or plant that
> is useful to man.

Sections 22 and 23 of the Act make a violation of Section 8 punishable by
a fine of up to five thousand dollars upon summary conviction for each offence. Each day the pollution continues is regarded as a separate offence.[123] Great reliance is placed on the ability of the federal and provincial governments to consummate joint agreements to designate water
quality areas and to establish water quality standards. Oil spill prevention and control is not particularly well suited to such an organizational
arrangement. The Canada Water Act deals with situations in which pollution is a deliberate violation, deserving criminal sanctions,[124] and makes
no attempt to establish regulations or initiate programs to prevent accidental pollution. In treating pollution as a criminal offence, the Act
makes no provision for compensating parties suffering damage from pollution. As a result, application of the Canada Water Act to oil pollution
problems offers nothing beyond those measures already present in other
legislation.

Canada Shipping Act

Prior to 1971 the Canada Shipping Act was largely limited to implementing international agreements. Originally passed in 1934, the
Act[125] was amended in 1956 to ratify and implement the 1954 International Convention. In the amendments Section 495A was added, giving the
Governor-in-Council power to make regulations to carry out provisions of
the 1954 Convention and authority to regulate and prevent the pollution of
oil from ships of any inland, minor, or other waters of Canada. Violations
of any of the regulations subjected the accused, upon summary conviction,
to a fine not exceeding five hundred dollars and/or imprisonment not exceeding six months.[126] In 1965 the Act was amended to implement the
1962 International Convention and the maximum fine was increased to a
sum not exceeding five thousand dollars.[127] Under 1969 amendments the
Minister was given discretion, where he has reasonable cause:[128]

> to believe that the cargo or fuel of a vessel that is in
> distress, stranded, wrecked, sunk or abandoned (a) is
> polluting or is likely to pollute any Canadian waters
> (b) constitutes or is likely to constitute a danger to
> waterfowl or marine life or (c) is damaging or is likely
> to damage coastal property or is interfering or is likely
> to interfere with the enjoyment thereof he may cause
> the vessel, its cargo or fuel to be destroyed or removed
> to such a place and sold in such manner, as he may
> direct.

Unlike the United States, Canada ratified the 1954 and 1962 Conventions
without reservation, but application of the Conventions in Canadian waters
has all the inherent limitations discussed in the second section of this chapter. In addition, the Act applies only to ships and does not cover all
sources of oil pollution, though, presumably, oil discharges from other
sources could be prosecuted under the Fisheries or Navigable Waters Protection Acts. The Canada Shipping Act has, however, been the most
widely used legislation to control oil spillage. The Marine Regulations

Branch of the Department of Transport is charged with supervising the Act and up to 1966 forty-five prosecutions were made, only two of which failed.[129] In enforcing the regulations the government has taken care to avoid prosecution under more than one act; however, the jail term provisions of the Canada Shipping Act are seldom invoked, and fines have not markedly reduced the number of oil pollution incidents. Perhaps the greatest failure of the Act prior to 1971, however, was a failure to deal with compensation, cleanup costs, and prevention.

Other laws have been passed by the federal parliament, but they either have limited spatial application, are redundant with other acts, or are so general that their application in combatting the threat of extraterritorial oil pollution is meaningless. Included in this category are the National Harbours Board Act, the National Parks Act and the Criminal Code.

Assessment of Federal Legislation to 1971

Canadian federal laws prior to 1971 ignored many of the issues which are fundamental to any system which internalizes all costs associated with the threat posed by the marine transport and drilling of oil. They focused on outmoded concepts of fines and criminal sanctions. Attachment of guilt or penal sanctions to polluters (while often satisfying to the public mind) is an unfair substitute for the equitable distribution of the social, ecological, and economic costs incurred when oil is spilled. Any workable system of internalizing costs must necessarily consider the extent and amount of liability, prevention and cleanup costs, methods of compensating victims, and the public's right to a pollution free environment. Moreover, until policies are formulated by the government to deal with these problems on a domestic scale, little can be done with threats of extraterritorial damage at the international level.

114

Skimmer system in operation after the Vanlene grounding

Photo: Howard Paish Assoc.

Health Act

Federal legislation, while often dated and inappropriate to current needs, is much more relevant than most provincial legislation. Early British Columbia legislation relating to water pollution was to be found in health legislation (and the regulations promulgated thereunder), and in laws relating to water use and appropriation. Many of these laws contained no specific reference to oil pollution, but they remain in the statutes. Under the Health Act Regulations:[130]

> No solid refuse or waste matter of any kind shall
> be deposited in any stream so as to obstruct its
> flow, or put into any stream or lake so as to
> pollute its waters, and no solid or liquid sewage
> matter from either public or private sewers shall
> be discharged into any stream or lake ... unless
> the best means have been first adopted to purify
> the same.

Persons violating the regulation were liable, upon summary conviction, for every such offence to a fine not exceeding one hundred dollars and/or imprisonment for six months. The Water Act forbids persons putting:[131]

> ... into any stream any sawdust, timber, tailing,
> gravel, refuse, carcass or other thing or substance
> after having been ordered by the Engineer or Water
> Recorder not to do so,

and every person violating this section is liable to a penalty not exceeding two hundred and fifty dollars and, in default of payment, to imprisonment not exceeding twelve months. Pollution has been subject to control by health and water authorities, and under such authorities, specific cases of pollution such as oil spillage are considered mainly from health or exploitation viewpoints. There has been a tendency to ignore broader questions of competitive water resource allocation and the social, ecological, and economic costs associated with competing uses. Jurisdictional confusion

is also apparent since no reference is made to tidal waters and the use of
the term "refuse" does not necessarily mean (as noted earlier in the discussion on federal legislation) that oil discharges on watercourses are specifically forbidden by the acts. Moreover, "pollution control (in the
sense of prior prevention) was difficult since the various provisions merely
made existing pollution the subject of penalties."[132]

Pollution Control Act

In 1956, after a dispute over discharge of sewage effluent, the
Provincial Legislative Assembly passed the Pollution Control Act.[133]
Under the Act a Pollution Control Board was established with power to determine what properties of water shall constitute a polluted condition, and
to set standards regarding the quality and character of the effluent which
may be discharged into all surface or ground waters of the province. Pollution was defined as "anything done, or any result or condition existing,
created, or likely to be created, affecting land or water which, in the
opinion of the Board, is detrimental to health, sanitation, or in the public
interest."[134] Discharge of waste into waters under the jurisdiction of the
province was prohibited unless the person first obtained a permit. Persons
whose rights might be affected could appeal the granting of any permit,
but the Board had sole discretion to make the objection the subject of a
hearing and to notify the objector of its decision. Individuals contravening any section of the Act or its regulations were subject to the same penalties that exist in the Water Act. Application of the Pollution Control
Act to oil pollution control was extremely limited. First, oil spillage
cannot be controlled by permit since incidents are most often not foreseen.
Second, the Act was originally intended to deal with municipalities and
municipal waste discharges and responsibility for the Act was given to the
Minister of Municipal Affairs.[135] Third, Section 2 placed a narrow definition on such terms as "effluent" and "works", further restricting the

Act to dealing with sewage outfall. Fourth, the province has little control over shipping, a primary source of oil pollution. Finally, the Pollution Control Board relied on the Health Branch of the Department of Health Services and Hospital Insurance for staff to enforce the Act - a further indication that the legislative assembly intended the Act to deal with water pollution caused by sewage discharge. The Pollution Control Act was amended in 1965, removing some of the restrictions present in the original act. Industrial wastes were specifically brought under the control of the Board. Administration of the Act was transferred from the Minister of Municipal Affairs to the Minister of Lands, Forests, and Water Resources. The government was given power, in accordance with the Civil Service Act, to hire employees whose specific duties were to enforce the Pollution Control Act. [136]

Pollution Control Act, 1967

In March, 1967 the 1965 Act was repealed and replaced by the Pollution Control Act, 1967. [137] The 1967 Act perpetuates the Pollution Control Board which is charged with determining polluted conditions, prescribing effluent standards, and appointing advisory and technical committees. In addition the Act provides for a Director of Pollution Control who is charged with administering the Act and is responsible for permit issue, amendment, and enforcement. Procedures for issuing permits, appeals, and decisions remain largely unchanged from the 1956 Act. Three interesting changes were made in the definition of "effluent," "pollution," and "waters" [138] which broaden the areas of the province that are covered and the number of substances considered to be pollutants, but there is no specific reference to oil. [139] In this respect British Columbia legislation regarding water pollution is similar to laws in the State of Washington prior to passage of the oil pollution acts in 1969, 1970 and 1971.

The Pollution Control Act of 1967 contained no penalty clauses, but amendments in 1968 re-established a penalty section to assist enforcement. Penalties are imposed on any person who discharges waste materials on or into any land or water without a permit. A polluter is liable, upon summary conviction, to a fine of up to one thousand dollars and/or imprisonment not exceeding three months. Continuing offences are subject to a fine of five hundred dollars for each day of commission. Thus the Pollution Control Act, 1967, as amended, absolutely prohibits the discharge of waste materials and establishes penalties as a means of enforcement. The right of prosecution, however, like that in the State of Washington under the Washington Oil Spill Act of 1970, lies solely with the Attorney-General acting on behalf of the Crown.

The Pollution Control Act of 1967, as amended, is the only statute that has even a partially meaningful, though indirect, bearing on oil pollution. It is so restrictive as to ignore liability, prevention facilities, cleanup costs, and compensation for areas of provincial jurisdiction; all essential features of laws to prevent and control oil pollution. Yet (as noted earlier) the provincial government is not prepared to pass specific legislation concerning oil pollution of the sea, even though the legislature authorized the provincial government to make agreements with the federal government to reduce the effects of pollution.[140] The Honorable Ray Williston, Minister of Lands, Forests, and Water Resources stated the policy of the provincial government:[141]

> ... it is important to avoid duplication and overlapping of administration, and the development of parallel organizations within the Federal and Provincial Governments.
>
> My staff is in contact with the federal officials and I understand that the Federal Shipping Act has been amended and further provisions are under consideration by the Federal Ministry of Transport to deal with oil spills and contingency plans. The Province will be advised when these measures have

119

been finalized. We will then be in a position to
consider supplementary measures if required from
the provincial point of view.

Under this policy, great reliance is placed on federal legislation to develop comprehensive oil pollution laws.

Statute Law Passed Since 1969

Arctic Waters Pollution Prevention Act

Since 1969 the federal parliament has enacted two laws specifically relating to oil pollution; one limited to the Canadian Arctic, and the
second to all other areas in Canada. The first law, the Arctic Waters
Pollution Prevention Act,[142] was passed by parliament in 1970. The Act
was introduced by the government in response to the voyage of the Manhattan through the Arctic ice pack in September, 1969. Reasons for this
introduction are complex; however, national pride, concern over Canadian
sovereignty in the Arctic archipelago, preservationist sentiment over the
ecological fragility of the area, and national security were all stimulated
by the Manhattan's voyage.[143] The government would have preferred an
international agreement (as opposed to unilateral action), but as Prime
Minister Trudeau noted:[144]

> It is well known that there is little or no environmental
> law on the international plane and that the law not
> in existence favours the interests of the shipping states
> and the shipping owners engaged in the large scale
> carriage of oil... There is an urgent need for the
> development of international law establishing that
> coastal states are entitled on the basis of fundamental
> principles of self defence, to protect their marine environ-
> ment and the living resources of the sea adjacent to
> their coasts.

The Act applies to Canadian waters north of the 60th parallel and
one hundred nautical miles seaward of the Canadian Arctic archipelago.
The discharge of wastes, including oil, are prohibited, and persons

discharging pollutants (including in the case of ships, both the ship owner and cargo owner) are liable, regardless of fault or negligence for incidents caused by third parties. Rigid safety requirements, evidence of financial responsibility, and reporting procedures are set forth in the Act. Fines of up to one hundred thousand dollars, seizure and forfeiture of both ship and cargo are provided for violators of these requirements.

While the Arctic Waters Pollution Prevention Act is not applicable to the study area, it demonstrated first, the concern of the federal government over oil pollution in areas over which it has jurisdiction. Second, passage of the Act also shows that the federal government has been willing in the past to take initiatives to prevent oil pollution, initiatives that are in advance of crisis and not entirely consistent with generally accepted international law. Such a policy could be particularly important in dealing with the oil pollution threat in Puget Sound and the Strait of Georgia, whether the policy manifests itself in stricter domestic legislation or, alternatively, in some form of international management with the United States. Finally, the Arctic Waters Pollution Prevention Act, along with the Canada Water Act, provides the basis for amendments to the Canada Shipping Act.

Canada Shipping Act

The 1971 amendments to the Canada Shipping Act[145] repealed the previous section on oil pollution and replaced it with Section 19. Under this section the discharge of pollutants is prohibited. "Pollutants" was defined previously in the Fisheries Act, and in the Arctic Waters Pollution Prevention Act. In the 1971 amendments, oil is specifically prohibited. The Governor-in-Council is given broad power to make regulations regarding discharge, to remove and destroy ships that are discharging or likely to discharge a pollutant, to establish the method of retention of oil or other waste of ships carrying pollutants, to make regulations

121

providing for the issue to the owner or master of a ship of a certificate sig-
nifying that the ship has complied with all the requirements of the Act, and
generally to establish rules for the operation of potential pollutant carrying
ships in Canadian waters. Areas of prohibited discharge include all Canadian waters south of the sixtieth parallel, those waters north of the sixtieth parallel that are not within the shipping safety control zone prescribed
under the Arctic Waters Pollution Prevention Act, and any fishing zone of
Canada designated under the Territorial Sea and Fishing Zones Act.

To enforce the regulations, the Act authorizes the Minister to appoint Pollution Prevention Officers. Such an officer is given power to inspect ships carrying pollutants, to order ships out of the waters to which
the Act applies (if he is satisfied that such an order is justified to prevent
discharge of a pollutant), and to regulate routes and speeds for ships carrying pollutants. Ship masters are required to give all Pollution Prevention
Officers all reasonable assistance.

Liability. Under the Act both the ship and cargo owner are jointly and
severally liable for all costs incurred by the government in repairing or
mitigating any damage to or destruction of life and property, and for all
actual loss or damage incurred by the government, a province or any other
person resulting from the discharge of a pollutant (such as oil) into Canadian waters. Liability does not depend upon proof of fault or negligence,
but no person is liable where he establishes discharge resulted from (1) the
conduct of other persons (2) an act of war, hostilities, civil war, insurrec-
tion, or a natural phenomenon of an exceptional, inevitable, and irrestible character, (3) an act or omission done with intent to cause damage
by a person other than any person for whose wrongful act or omission he is
by law responsible and (4) the negligence or wrongful act of any person
or government in the installation or maintenance of lights or other navigational aids. Where the incident occurs without actual fault or privity on

the part of the ship or cargo owner, liability is limited to 2000 gold francs for each ton of the ship's tonnage (approximately 134 dollars per ton) or 210,000,000 gold francs (approximately 14 million dollars) whichever is less. However, where the incident occurs with fault or privity, there is unlimited liability. All claims may be sued for and recovered in Admiralty Court.

Maritime Pollution Claims Fund. The Act takes initial steps to provide victims of oil spillage with compensation through a new Maritime Pollution Claims Fund, and provides for an Administrator to head the fund. Financing for the fund is based on a tax of up to fifteen cents per ton for each ton of oil imported by ship into Canada as bulk cargo, and each ton of oil shipped from any place in Canada in bulk as cargo of a ship; the actual tax being regulated by the Governor-in-Council. For the purpose of the Act, barges are classified the same as ships. The Administrator is empowered to review all claims made under the Act and to direct payment of all settlements and judgements against the Fund. When a claim by a victim of oil spillage is not fully met by the amount of liability available, the Administrator can direct payment out of the Fund to satisfy the remainder of the claim. This is usually done after a court has rendered a decision as to the total amount of the claim. Moreover, when ships causing pollution cannot be identified, the Fund can be held liable for the damages incurred. Proceedings in such instances are to be instituted in Admiralty Court.

Fishermen are given special consideration under Section 755 of the Act. When a fisherman alleges that he has suffered a loss of income (including future income but not assets) from his activities as a fisherman, resulting from a discharge of a pollutant that was caused by a ship and that is not recoverable under any law, he may give notice in writing to the Administrator. The Administrator, in turn, will direct payment out of the Fund or appoint an assessor to consider the case. The assessor is empowered to rule on the alleged loss of income and set the amount of loss.

Priorities are established for paying money out of the Fund. Costs and expenses incurred by the Administrator have first claims against the Fund. These expenses are followed in order by claims of fishermen for loss of income, actual losses or damages, and expenses incidental to taking action against actual or potential polluters.

Penalties. In addition to the compensation section, the Act also establishes substantial penalties as a means of enforcement. Persons or ships that discharge a pollutant in contravention of any regulations are guilty of an offence, and liable, upon summary conviction, to a fine not exceeding one hundred thousand dollars. Smaller fines are prescribed for persons attempting to avoid payments into the Fund. In addition, when a Pollution Prevention Officer has reasonable cause to believe that the regulations of the Act have been contravened, he may, with the consent of the Minister, seize the ship.

The Canada Shipping Act and the Water Quality Improvement Act

The amendments to the Canada Shipping Act provide Canada with some of the strictest domestic oil pollution laws. These amendments also provide a more equitable distribution of the costs and risks involved with oil transport than does the Water Quality Improvement Act of 1970. However, neither of the acts is definitive. The Canadian law holds both the ship owner and the cargo owner liable, while the Water Quality Improvement Act of 1970 holds only the owner or operator of the ship liable. A Maritime Pollution Claims Fund is created, which, on a tax of fifteen cents per ton for all oil moving by water, is expected to yield three million dollars annually. This amount does not approach the 35 million dollars appropriated in the United States, but, under Canadian law, a method is established to replenish the Fund when claims are made against it.

124

Difficulties may arise should a major oil spill occur in the first few years of operation, as no provisions are made for the Fund to go into debt. By establishing such a fund, the Canadian parliament avoids the possibility of not collecting where the cleanup costs amount to more than 14 million dollars (as is now the case with United States federal law). Moreover, the Canada Shipping Act amendments provide for claims by provinces and individuals against the amount of owner liability and the Maritime Pollution Claims Fund. In the United States claims are limited to government expenses. Perhaps the most significant aspect of the Canada Shipping Act amendments is the move to compensate some victims of oil spillage for loss of present and future income, a principle which the Water Quality Improvement Act fails to even consider. Attention is also given in the Canada Shipping Act to partial control of such facilities as docks. In this respect, the federal government has powers similar to (though not as extensive as) those existing in the State of Maine. While primarily domestic legislation, the amendments to the Canada Shipping Act anticipate the possibility of more comprehensive international accord on oil pollution. Liability is quoted in francs, since this currency is used in most international conventions on oil pollution. Through the extensive regulations, the government is also given power to implement amendments to the 1969 Brussels Conventions (and other, similar conventions) without having to obtain specific parliamentary approval in the form of implementing legislation.

In two areas the Canada Shipping Act amendments are weaker than the Water Quality Improvement Act. First, there is no legislative mandate requiring preparation of national and regional contingency plans (although the government has begun to prepare plans by administrative decision). Second, the Act is applicable only to oil pollution emanating from ships, and does not consider pollution that may result from offshore drilling.

Assessment of the Canada Shipping Act

The amendments, like much legislation, have inherent weaknesses.

The law cannot be applied to foreign ships heading for foreign ports unless the ships pass through Canadian waters. Thus the Act will be of little consequence to the proposed shipment of oil from Alaska to Puget Sound if the tankers remain outside Canadian waters; although, as noted earlier, the legislation would be important in any claim for extraterritorial damage which might come before an international court.

The Canada Shipping Act is not the only federal legislation dealing with oil pollution. Control over the problem is shared by several departments: Environment, Public Works, Indian Affairs and Northern Development, External Affairs, and Transport. The problem of oil pollution calls for a good deal of cooperation, but Canada lacks an independent agency or even a permanent interdepartmental committee to deal with the problem. Some of this malaise may be overcome through the Department of the Environment but, even with this department, oil pollution control will still be the domain of several ministers.

Several improvements might be made to the legislation to minimize discharge and to equitably distribute the risks and costs associated with drilling and transport of oil. Compensation should be extended to persons suffering direct damage in addition to the claims of fishermen for income loss and also to persons suffering indirect damage. Greater emphasis should be placed on prevention through such facilities as radar control of ships. Such emphasis could result from administrative decision or legislative action. Liability should be extended to oil pollution resulting from drilling operations, and claims should be able to be made against the Fund. Finally, as discussed earlier in part two of this chapter, more consideration should be given to shifting the burden of proof to the polluter.

Contingency Plans

Contingency plans in Canada are still being developed and are not as far advanced as those in the United States. Unlike the United

126

States Congress, the Canadian parliament has not passed any legislation requiring the government to prepare and publish a national contingency plan to combat oil pollution. The government recognized the need for an organized response to such incidents after an ad hoc arrangement was used to deal with the oil escaping from the tanker Arrow off the coast of Nova Scotia. Discussions were held between officials of the Department of Transport, of other federal departments, and of the Task Force supervising cleanup of the Arrow spill. As a result, the government, by administrative decision, prepared in July, 1970 the "Interim Federal Contingency Plan for Combatting Oil and Toxic Material Spills"[146] for review by interested parties. The Interim Plan is, at present, the only official document available and in force.

Interim Plan

While the Contingency Plan is of interim status, it provides the framework within which the final National Contingency Plan will operate. The Plan provides for a federal response to major oil spills in areas of federal jurisdiction. On-Scene Co-Ordinators are named for the Victoria, Vancouver, and Prince Rupert Sub-Regions on the West Coast. (The Victoria and Vancouver Sub-Regions encompass the waters of the study area). The federal plan intends that provincial agencies participate, especially to clean up small spills in areas of provincial jurisdiction. The level of provincial participation will in turn influence the degree of federal response. Should the provincial and local governments fail to acquire materials, the federal government is prepared to organize regional capabilities and acquire equipment to fight local spills, but the federal plan recognizes that participation of the provinces "is not only desirable but essential."[147] Provincial and local participation will only come after conclusion of a federal-provincial agreement. These agreements, as indicated earlier, will be difficult to consummate in absence of a crisis, especially in view of British Columbia's stated intention to let the federal government prepare plans for inland waters and introduce supplementary measures only if federal plans prove inadequate. Without provincial participation and cooperation, operational capability of the plan will be considerably delayed.

Persons discharging oil are under no legal obligation to report spills, but the Canada Shipping Act provides that they shall be responsible for all cleanup costs. When reports are made of spills occurring in waters of federal jurisdiction, they are forwarded to the On-Scene Co-Ordinators. The Co-Ordinator is to assess the situation, utilize local equipment to combat spills, and, if necessary, seek federal assistance by informing (in the case of a west coast spill) the Regional Co-Ordinator for the West Coast. The On-Scene Co-Ordinators are provided with manuals indicating the latest technology available in cleaning up oil spills. If the discharge is beyond the capability of the region, the On-Scene Co-Ordinators can call on the Inter-Departmental Committee on Contingency Planning in Ottawa for physical and technical aid. The Committee, in turn, can then authorize a federal response, and the Departments of National Defence and Transport are charged with overseeing the operation.

Beyond the skeleton framework, meetings have been held on the west coast to supplement the Interim Plan. Peat moss stocks have been established, small stocks of emulsifiers are now carried at marine bases, and critical water areas have been designated. Inventories have been made of local firms having equipment and experience to effectively clean up oil spills. Oil companies have agreed to identify all major oil shipments to and from the west coast on a 24 hour basis by telex, giving the name of the vessel, routing, quantity, and type of oil being carried. In addition many of the oil companies have made provisions on their ships to clean up small spills, and have arranged with local firms to collect materials which could be used for containing larger spills.

Cooperation between Canadian and United States Contingency Plans

There is a clear recognition in the Contingency Plan that close contacts should be made with the United States to coordinate contingency plans but, as yet, little has been achieved. In discussing the close cooperation maintained with the United States in the Great Lakes Region, the Interim Contingency Plan comments:[148]

> International co-ordination for the connecting channels of the Great Lakes has been established through the International Joint Commission. This could be expanded to cover the lakes, although it must be appreciated that IJC is not an operating body.

Unofficial discussions about the need for close coordination have been held between Canada and the United States for specific areas such as the west coast. Talks have been held, too, at the senior civil servant and ministerial levels in Ottawa and Washington, but there have been no concrete announcements regarding either contingency plan coordination or a role for the International Joint Commission on the Pacific Coast.

Assessment of the Contingency Plans

Contingency plans in Canada mirror the malaise evident in legislation regarding oil pollution. Neither the provincial nor the federal government feels it has the responsibility of cleaning up all oil that is spilled along the coast. The provincial government seems willing to let federal authorities undertake initial preparations, but the federal government has no guarantee that the provincial authorities will not initiate conflicting and overlapping plans. The federal government, on the other hand, acknowledges responsibility for cleaning up spills in waters of federal jurisdiction, but does not assume responsibility for spills in areas of provincial jurisdiction. As noted earlier the jurisdiction over offshore and inland waters remains confused. To resolve this dispute a federal-provincial agreement is needed, yet there are several sources of disagreement

129

between the province and the federal government between the province
and the federal government, and the prevention and control of oil pollution is often ignored within these larger conflicts. If there is to be any
formal arrangement with the United States concerning contingency plans,
regional prevention measures, and possible extraterritorial compensation,
a federal-provincial agreement will be a fundamental prerequisite to negotiation. Neither the United States nor the Canadian government would
be willing to suffer delays and uncertainties like those resulting from the
Columbia River dispute, where firm federal-provincial agreement was
lacking. Unless the federal and provincial governments resolve their jurisdictional dispute through a joint agreement the Canada-British Columbia
response to the threat of extraterritorial damage from oil pollution in Puget Sound and the Strait of Georgia will remain fragmented and impotent.

AN EVALUATION OF THE RESPONSES

Puget Sound and the Strait of Georgia possess unique characteristics as a salt water estuarine area. Scientists have argued that major oil
spills, especially those capable of causing extraterritorial damage, present
a distinct threat to the ecological balance of the region. An earlier section of this chapter argued that the existing amount of petroleum trade in
the area was capable of severely polluting waters in the estuary and that
prospects of oil deliveries from Alaska would only increase the existing
threat.

Citizens in North America have been seeking more amenities in
life through such pursuits as outdoor recreation while at the same time desiring a better standard of living. On the one hand the public (on both
sides of the Canada-United States border) desires to maintain the quality
of the marine and coastal environments in the study area. Yet, on the
other hand, it needs (under present technology) petroleum to fuel the industrial and affluent society and maintain an advanced standard of living.

The transportation of oil in marine areas is not completely incompatible with the maintenance of a quality environment, but, in using the oceans without adequate means of prevention, control, and compensation, we are threatening to lower water quality below an ecologically acceptable level and to create an inefficient allocation of the resource.

Utilization of any resource requires consideration of the full range of consequences emanating from exploitation. Often the level of one use must be reduced, controlled, or curtailed because it interferes or conflicts with other uses of the resource. The problems posed by this study address themselves to the need for tradeoffs associated with the use of the salt water estuary comprised of the Strait of Juan de Fuca, Puget Sound, and the Strait of Georgia. The above examination of the United States-Washington and Canada-British Columbia response to the threat posed by using the marine environment for oil drilling and transportation revealed that: (1) little consideration has been given to the public's right to a pollution free environment; (2) laws have focused on prohibitive discharges, sanctions, limited liability, and penalties; (3) compensation for victims suffering damage from spilled oil is difficult to obtain and is largely limited to claims of direct damage; (4) contingency and prevention plans are still being developed and are restricted to domestic responses; and (5) no system exists, at either the domestic or international level, to internalize the full range of costs associated with the threat. Some domestic laws contain principles that could be incorporated into future domestic legislation and international arrangements designed to internalize extraterritorial damage from major oil spills. In many instances the principles and powers of existing laws are sufficient to reduce potential damage, but these laws do not account for all costs and governments have not rigidly enforced their provisions. In other cases new laws and institutions are needed to meet the problems of damage from major oil spills.

The Legality of Oil Discharges

The discharge of oil into or upon the navigable waters or territorial seas is illegal (with minor exceptions) in the United States and in waters under federal jurisdiction in Canada. Difficulties might arise, for instance, if oil spilled on land from a pipeline entered water under provincial jurisdiction and then spread to Canadian federal waters and then to United States waters. Under British Columbia law, the original discharge would not be illegal. The federal government would be involved because of the international damages, but might not be able to prosecute the violation under federal statutes because the spill occurred in an area of provincial jurisdiction. Conversely, if the spill occurred in United States waters and damaged only provincial land and waters, the province might not be in a strong position to collect damages. If precedents in the Trail Smelter case were applied, the laws of the jurisdiction damaged would be employed by an international tribunal in awarding damages. Since the discharge of oil is not specifically prohibited under British Columbia law, it is possible that damages might not be awarded. There is a need therefore, in the absence of a firm international agreement between the United States and Canada, for British Columbia to prohibit the discharge of oil upon the land and waters under its jurisdiction.

Rights to a Pollution Free Environment

In the two major federal statutes (the Water Quality Improvement Act of 1970 and the Canada Shipping Act of 1971) there is an implicit recognition that the public is entitled to a marine environment free from oil, but this recognition is not made explicit in the statutes. The Washington Oil Spill Act of 1970 recognizes the rights of Washington citizens to pure and abundant waters, but authorizes only the Attorney-General to take action on the public's behalf. British Columbia lacks any statutes defining the rights of its citizens to a pollution free environment. If this

132

right were made explicit (either by statute or constitutional revision), individuals might have greater success in private actions and private citizens
might then be able to take action on the public's behalf for damages against
public property. In addition, uniformity in the public's right to a pollution free environment in all jurisdictions present in the study area would
enhance any claims for extraterritorial damages.[149]

Extraterritorial Oil Damage

Discharges are prohibited in the territorial waters of Canada and
the United States and in waters designated in the 1962 International Convention. In the territorial waters, prosecutions have been made for violation of domestic legislation, but there have been no major incidents resulting in claims for extraterritorial damage emanating from another jurisdiction. Several incidents have occurred on the high seas causing damage
to coastal areas; however, existing international laws favor the polluter
to the detriment of the coastal state. Detection of oil polluters is difficult because of the vast areas of the oceans that have to be policed by
visual surveillance. In addition, if a polluter is caught discharging oil
outside of territorial waters but inside the zones established by the 1962
Convention, prosecution is left to the flag state. Such an arrangement
provides inadequate protection against pollution because control is vested
in the state having the least interest in anti-pollution measures. Governments, especially those having control of large fleets, have been miserly
in their willingness to surrender a measure of their sovereignty over their
own flag ships, either to international organizations or to coastal states
who stand to suffer most when oil is discharged at sea. If coastal governments were given authority through international convention or, failing
that, through their own domestic legislation to expand their contiguous
zones and to take action against a source of pollution at sea, pollution
from tankers might be reduced and coastal states better enabled to

133

safeguard their shores and territorial waters.[150]

The Areal Scale of Control

Preventive measures in the study area are oriented to domestic needs and do not reflect the regional character of marine oil pollution. Both Canada and the United States provide aids to navigation, charts to show navigable routes, and rules to govern ship operations in their own territorial waters. There is, however, no traffic control, routing, or lane separation organized on a regional basis. As a result, ships, while under the control of pilots, operate by the rules established in each country. Such pilots have no control over other ships on either side of the border. With the number of ships increasing, the potential for collision grows; yet the methods of ship control remain antiquated. Similarly, as noted above, contingency plans are being developed within a domestic framework without regard to regional needs. Preventive measures of pollution control which fail to reflect the regional nature of the oil pollution problem will not be sufficient to reduce the threat of extraterritorial damage. Thus it is imperative that the informal consultations between Canada and the United States be translated into firm arrangements to control ship traffic and organize contingency plans and navigational aids on a regional basis for Puget Sound and the Strait of Georgia.

Possible Unilateral Actions

Ideally, other preventive measures, such as standards for ship construction and navigational competence, are better left to international conventions. A multitude of conflicting regulations for vessels engaged in world commerce would severely hamper trade. Yet the pace of international compromise and agreement is slow, and coastal nations who feel oil pollution poses an immediate danger are likely to find national solutions attractive. The Canada Shipping Act, for example, authorizes the Governor-in-Council to make regulations regarding the construction of

134

ships carrying hazardous materials in Canadian waters. If the regulations
of the coastal state and the flag state are dissimilar, the ship could be prevented from entering Canadian waters. Unilateral action can impede international efforts at cooperation and coordination, but such action can
also help achieve higher standards of ship construction. If the nation were
a major trading state, unilateral action could form the basis for speedier
international accord, if not by convention, then by acquiescence on the
part of the ship owners. No conclusive case can be made for unilateral
as opposed to international action to achieve stricter standards and regulations, but, if existing international standards do not reduce the threat of
oil pollution, any nation would be justified in promulgating domestic standards to protect its own shores. If the preventive measures discussed above
are to be organized on a regional basis, any decision on standards for ship
construction and navigational competence will require consultation and
agreement between Canada and the United States. Stricter bilateral regulations between two major trading nations in turn may lead more quickly
to better multilateral standards for preventing oil pollution.

Any system which attempts to account for all potential costs involved in drilling for and transporting oil must necessarily include provisions for dealing with damages that result when oil is spilled, even if that
damage does not occur initially in the waters of a particular state. Hovanesian states the case:[151]

> It would seem... clear that supertankers present
> supranational problems which simply cannot be solved
> by traditional means. A chance foundering cannot
> be legislated out of existence by the government
> of a single nation. Therefore, recognizing that
> the phenomenon is not to be stopped by words, the
> least which should be prudently done is to set
> up a mechanism which is designed to afford swift
> and comprehensive redress to those whose land and
> possessions are engulfed. In the face of such a
> calamity, private parties may be unable to help

themselves, nor seek effective legal redress in
distant and costly courts of law. At the same
time, the government of even the poorest nations
should be encouraged to give every assistance
within their means to their affected citizens.

Private remedies have been shown to be singularly ineffective as
a means of compensating victims of oil spillage. Actions of nuisance
would be more successful if the courts determined, because of statute law,
that the carriage of oil was an ultrahazardous activity. Nuisance has usually been interpreted as injury related to the enjoyment of property rights
and not as damage to property. Such an interpretation may give rise to
liability in the absence of negligence and, therefore, broaden the chances
of recovery for direct damage to property. All private actions, but especially negligence claims, would be greatly enhanced if the burden of proof
were shifted to the polluter to prove he has not deviated from a particular
standard of care. Canada and Washington State have incorporated this
concept into their statutes by introducing the concept of no fault liability, but the United States and British Columbia lack specific statute law
on this point. Even with this broadening, the chances of victims recov-
ering indirect damages are not improved. Judgements which do not take
account of indirect damages are inequitable and result in a misallocation
of resources. In cases of extraterritorial damage in the study area, changes
in Washington and Canadian law will mean little unless a state to state
claim is launched between Canada and the United States. The chances
of such an action being agreed upon are now possibly even more remote
as a result of the 1970 Canadian decision to expressly remove disputes involving marine pollution matters from the jurisdiction of the International
Court of Justice.[152] In the absence of an international court or bilateral
convention agreeing to litigation, the only alternative is a special tribunal similar to that convened in the Trail Smelter dispute. Since the relevant law used in that dispute was that of the territory damaged, it is

imperative that each government jurisdiction operating in the study area has domestic laws guaranteeing full compensation for all costs suffered in oil spills. Then, if a claim for extraterritorial damage were made and a tribunal established, each citizen could expect to recover all damages to his property. Without comprehensive domestic laws, claims for extraterritorial damage are likely to fail.

Comprehensive domestic laws must also provide adequate rules of liability for persons engaged in drilling for and transporting oil. The Washington Oil Spill Act of 1970 is by far the strictest, providing for unlimited liability for those who own or control the spilled oil. United States federal law is more restrictive. If the discharge was not caused by wilful negligence, the liability of the owner or operator of the vessel is limited to 100 dollars per gross registered ton or 14 million dollars whichever is less. Offshore facilities are limited to 8 million dollars liability, and only the government can collect costs from this liability. Canadian law is similar to United States federal law with five major exceptions: (1) the liability limit is 134 dollars per ton or 14 million dollars whichever is less; (2) ship and cargo owners are jointly and severally liable; (3) claims are not limited to government cleanup costs; (4) there is no mention of liability for discharges from offshore or onshore facilities; and (5) a compensation fund is established by the Canada Shipping Act for discharges emanating from unknown sources and for claims above the 14 million dollar limit.

A Model Domestic Law

The Water Quality Improvement Act of 1970, the Washington Oil Spill Act of 1970, and the Canada Shipping Act contain provisions which might be incorporated into a model domestic law to ensure adequate compensation in the case of oil spill damage. First owners of oil and operators of vessels and onshore and offshore facilities should be jointly and severally liable for all damage. Second, all claims, government and individual, direct and indirect, should be satisfied. Third, an adequate amount of liability should be available to satisfy all claims. Some ship and cargo owners have argued that there should be a ceiling placed on the amount of liability so they can insure themselves against oil spillage. They maintain owners would be unable to obtain insurance if there were no ceiling on liability. Others contend that airlines have been able to insure themselves even though there is no limit on liability, and that, if the owners cannot absorb such risks and remain competitive, they should not be in business.[153] The 1969 Conventions propose limits that were adopted in the Canada Shipping Act. To facilitate international accord, these amounts could be incorporated into domestic legislation. It is clear, however, that beyond any limit of liability, governments should establish compensation funds to finance oil spill prevention measures and to compensate those suffering oil spill damage when the polluter is not known. Such a compensation fund could be based on the amount of oil tonnage entering or leaving domestic ports. An upper limit could be set for the fund, depending on the threat of damage and the rate of tax adjusted accordingly.

In Chapter Two the arguments of Brown, Mar, Parker, Crutchfield, and Kneese were used to demonstrate that the most efficient technique by which oil pollution could be controlled is by forcing the polluter to internalize, through a system of effluent charges based on a non-linear damage function, the technological externalities created by his oil discharge. Because oil discharges are not usually deliberate and frequent acts occurring at specific points, the polluter cannot always be identified. Nor can the amount of oil spilled be controlled, so the type of charges envisaged by Brown and Mar are not directly applicable. Nevertheless, establishment of controlled liability and creation of a compensation fund would be a form of indirect charge. If the tax could be adjusted to such a level that the fund was sufficiently large to absorb the damages caused by some future catastrophe, the tax might achieve the same goals desired by Brown and Mar. The primary difficulty will rest with adjusting the liability and tax so that they accurately reflect potential damage costs and cause the potential polluter to take preventive actions to minimize the costs of spillage. Moreover, if the liability and tax could be applied on a regional basis for Puget Sound and the Strait of Georgia there would be a means to internalize any external damage created by spilled oil.

Conclusion

This chapter has demonstrated that the response to the threat of extraterritorial oil pollution in Puget Sound and the Strait of Georgia has been decidedly domestic. Governments have made attempts to meet the demands posed by domestic spills, but these laws would have international application only in the convening of an international tribunal. Little consideration has been given to the problems involved in compensating victims of large oil spills or to the regional nature of the marine pollution problem. National and subnational institutions seem fractured and incapable of internalizing the externalities associated with a major oil spill. There are, however, a number of international plans and bilateral conventions which might be used to reduce the impact of major oil spills in Puget Sound and the Strait of Georgia. Chapter Four considers these approaches and assesses their ability to prevent oil spillage and minimize damage.

REFERENCES

1. Battelle Memorial Institute, Pacific Northwest Laboratories, Oil Spillage Study: Literature Search and Critical Evaluation for Selection of Promising Techniques to Control and Prevent Damage; to Department of Transportation, United States Coast Guard, Washington, D.C. Washington: United States Department of Commerce, Clearinghouse for Federal Scientific and Technical Information (1967), pp. 2-8, 4-44, 4-45.

2. "Oil Spill Could Foul B.C. Waters," Vancouver Sun, January 29, 1971, p. 2. For a description of tidal flows in Puget Sound waters see Puget Sound Task Force, Comprehensive Water Resources Study. Navigation: Puget Sound and Adjacent Waters. (1970), preliminary draft, Appendix VIII.

3. GLUDE, J. B. "Information Requirements for Rational Decision-Making in the Control of Coastal and Estuarine Oil Pollution," paper given to the FAO Technical Conference on Marine Pollution and its Effects on Living Resources and Fishing, Rome, Italy, December 9-18, 1970. For a more comprehensive summary of the specific environmental problems in Puget Sound and particularly the Strait of Georgia, see Canada, Parliament, House of Commons. Order 209 (November 12, 1971), and Aide-Memoire to U.S.A. Government (August 18, 1971) and attachments.

4. Trans Mountain Oil Pipe Line Company. General Article. Vancouver: Trans Mountain Oil Pipe Line Company (1969), Supplemented (1970).

5. HACKING, N. "Oil Moves Out by Sea," Vancouver Province, October 23, 1970, p. 23.

6. Personal communication, Edward Weymouth, Executive Secretary, Western Oil and Gas Association, Seattle (January 19, 1971). See also Annual Report, Marine Exchange, Seattle Chamber of Commerce for the number of oil tanker sailings and arrivals and Canada, Parliament, House of Commons, Special Committee on Environmental Pollution. Proceedings and Third Report, Issue 21.

Ottawa: Queen's Printer (1971), p. 14.

7. Personal communication, Sidney S. Martin, Marine Manager, Imperial Oil Limited, Vancouver, (January 28, 1971).

8. BURT, L. "Cole Will Turn Down Oil Drilling Permits," Seattle Times, November 16, 1970, p. 1; "Oil Drillers Attorney Won't Appeal State Ban," Seattle Times, November 17, 1970, p. C 17; "Cole Rules Against Drilling in Sound," Seattle Post-Intelligencer, November 17, 1970, p. 1 and other articles on page 10 of the same edition; WILLIAMS, H. "Puget Sound: a Flood Tide for Aquaculture?," Seattle Times, February 7, 1971, p. A 15.

9. CAPLAN, N. "Offshore Mineral Rights: Anatomy of a Federal-Provincial Conflict," Journal of Canadian Studies, No. 5 (February 1970), pp. 50-61. See also LITTLE, C. H. "Offshore Exploration for Gas and Oil," Canadian Geographical Journal, No. 77 (October 1968), pp. 108-115.

10. McCONNELL, B. "Will Bennett Become Ruler of the Strait?," Vancouver Province, January 14, 1971.

11. See comments in the Vancouver Province, December 6, 1969 and Vancouver Sun, December 17, 1969. In 1970 the Science Council of Canada recommended a federal-provincial feasibility study of oil drilling in the Strait of Georgia but this was summarily rejected by Davis. See Science Council of Canada. Canada Science and the Oceans. Ottawa: Science Council of Canada (1970); Vancouver Sun, November 25, 1970, p. 71; and DAVIS, J., Minister of Fisheries and Forestry. Marine Parks for More People. Address to Save our Parkland Association, Vancouver (November 6, 1970).

12. See State of Washington, Department of Ecology. Reported Oil and Hazardous Material Spills in Washington State. (Monthly since January 1970).

13. "Oil and Water: The imagination pales...," Vancouver Province, January 20, 1970, p. 3.

14. National Energy Board. Energy Supply and Demand in Canada

and Export Demand for Canadian Energy, 1966 to 1990.
Ottawa: Queen's Printer (1969), p. 144.

15. Puget Sound Task Force, op. cit., p. 2-33.

16. Ibid., p. 2-33, 2-34.

17. Honeywell, Marine Systems Center. A Proposed Automated Marine Traffic Advisory System for Puget Sound. Seattle
(November 6, 1970), 21 pages and appendices.
Industry representatives have discounted this assertion
by Honeywell and point to the safety record of ships
on Puget Sound and programs underway to reduce the
possibility of spillage. See Seattle Chamber of Commerce. "Ship Movements and Petroleum Transportation
on Puget Sound Area, State of Washington, White
Paper." (April 1971, 11 pages).

18. CRUTCHFIELD, J. A., BISH, R. L., MALENG, N. K., and
WARREN, R. O. Socioeconomic, Institutional and
Legal Considerations in the Management of Puget Sound,
Final Report submitted to the Federal Water Pollution
Control Administration, Contract 14-12-420. (Au-
gust 15, 1969), manuscript copy, p. 175.

19. ROGERS, G. W. "Change in Alaska: The 1960's and After," in
ROGERS, G. W. (ed.) Change in Alaska – People,
Petroleum and Politics. College, Alaska: University of
Alaska Press (1970), pp. 10-12.

20. CRUTCHFIELD, et al., op. cit., p. 176.

21. Information for this paragraph comes from the Subcommittee on
Oil Spill Taxation, Oceanographic Institute of Washington, "Minutes of Meeting," December 9, 1970.
(Quoted from remarks by Emergy A. Winkler, Atlantic
Richfield Company, Philadelphia). See also Alyeska's
(the pipeline company formed to build the Trans Alaska
Pipeline) reply to a Canadian diplomatic note delivered to the United States on August 18, 1971, ("U.S.
Oil Firms Rap Ottawa," Vancouver Province, December 22, 1971, p. 8). In their reply Alyeska maintains
that less than 10 percent of the total tanker traffic from
the Alaska pipeline is projected to move to Puget Sound.

In 1975, the projected starting date, an estimated six
tankers per month would enter Puget Sound or about
70 tankers per year. Nine years later this would increase to only 80 or fewer tankers per year.

Dean maintains that the major markets for Alaskan
oil will be in California. See DEAN, C. J. "Markets
for Alaskan oil," Scottish Geographical Magazine,
No. 87 (September 1971), pp. 147-150. Prime Minister Sato of Japan has expressed interest in utilizing
Alaska oil in Japan. See "Japan Hopes to Buy Oil
from North Slope," Seattle Times, January 8, 1972,
p. 1. There is, however, little documentary evidence
to substantiate these claims in the form of charters or
tanker orders other than the plans of Atlantic Richfield
noted above.

22.   University of California, Davis. Legal Control of Water Pollution,
       Vol. 1. Davis: University of California, Davis, Law
       Review (1969), p. 168. See also SHUTLER, N. D.
       "Pollution of the Sea by Oil," Houston Law Review,
       No. 7 (March 1970), p. 423. "The recognized rights
       of innocent passage of a foreign vessel in the territor-
       ial sea is qualified only by its duty to comply with the
       sovereign's regulations while in that sea. However, a
       nation can take no action against a vessel for a crime
       committed before it enters its territorial waters if it is
       only passing through."

23.   L. H. J. Legault of the Canadian Department of External Affairs,
       Ottawa, contends that the exercise of reasonable con-
       trol of a State over foreign vessels in areas of the high
       sea contiguous to its territorial waters does not violate
       international law. He feels certain elements of the
       rule of reasonableness must be made clear. First, the
       rule does not depend on the acquiescence of another
       state before it can be applied. Second, there is no
       limit on the distance to which a State might assert pro-
       tective control. Third, vessels do not have to be bound
       for the nation exercising unilateral action. Fourth, the
       rule would appear to encompass punitive as well as pre-
       ventive action. Legault's argument would appear to be
       the rationale used by the Canadian Government in
       adopting the Arctic Waters Pollution Prevention Act;
       however, this view is not widely held and its validity

143

in international law is questionable. See LEGAULT,
L. H. J. "The Freedom of the Seas: A License to
Pollute?," paper given at the Symposium on International Legal Problems of Pollution sponsored by the Canadian Branch of the International Law Association,
the University of British Columbia, and the Department
of External Affairs, Ottawa and held in Vancouver
(September 8-11, 1970), pp. 14-15.

24.   Three types of international spills are possible under these conditions. The first would arise along the international
boundary when oil spilled in the Strait of Georgia, for
example, caused damage in Puget Sound. The second
could result when oil spilled in the territorial waters
of the United States near Cape Flattery in the Strait of
Juan de Fuca spread beyond the contiguous zone to the
high seas and then re-entered the Strait of Juan de Fuca
and damaged the shoreline or marine life in Canada.
Responses to this type of spill are discussed later in this
chapter. Both of these types are different from a spill
which occurs on the high seas and then affects domestic
waters because the spill would not have occurred first
in a sovereign jurisdiction. This type of spill is discussed in Chapter Four.

25.   ULLMAN, E. L. "Political Geography in the Pacific Northwest,"
Scottish Geographical Magazine, No. 54 (July 1938),
pp. 236-239.

26.   O'CONNELL, D. M. "Continental Shelf Oil Disasters: Challenge to International Pollution Control," Cornell Law
Review, No. 55 (November 1969), p. 126.

27.   University of California, Davis, op. cit., pp. 171-173.

28.   LESTER, A. P. "River Pollution in International Law," American
Journal of International Law, No. 57 (1963), pp. 833-
834.

29.   READ, J. E. "The Trail Smelter Dispute," Canadian Yearbook of
International Law (1963), pp. 222-227.

30.   KUHN, A. K. "The Trail Smelter Arbitration - United States and
Canada," American Journal of International Law,
No. 32 (1938), p. 787.

31.   LESTER, op. cit., pp. 839-840.

32.   LEGAULT, op. cit., p. 11.

33.   State of Washington, Department of Ecology. Laws and Oil Spill
         Emergency Procedures. Olympia: Department of
         Ecology (1971), p. 2.

34.   Personal communication, James C. Willman, Chief, Oil Pollution
         and Hazardous Materials, Environmental Protection
         Agency, Northwest Region (March 10, 1971).

35.   Ibid.

36.   SHUTLER, op. cit., p. 423.

37.   SNOKE, K. P. "Admirality Law: California Sues a Vessel in
         Rem for Oil Discharge Damages to its Water and Ma-
         rine Life," Tulsa Law Journal, No. 6 (August 1970),
         pp. 257-273.

38.   "Oil Pollution of the Sea," Harvard International Law Journal,
         No. 10 (Spring 1969), p. 347.

39.   SWEENEY, J. C. "Oil Pollution of the Oceans," Fordham Law
         Review, No. 37 (December 1968), p. 166.

40.   SINGLETON, J. F. "Pollution of the Marine Environment from
         Outer Continental Shelf Operations," South Carolina
         Law Review, No. 22 (Spring 1970), pp. 237-238.

41.   "Oil Pollution of the Sea," Harvard International Law Journal,
         op. cit., p. 348.

42.   SINGLETON, op. cit., p. 238.

43.   EDWARDS, M. N. "The Role of the Federal Government in Con-
         trolling Oil Pollution at Sea," in HOULT, D. P. (ed.)
         Oil on the Sea. New York: Plenum Press (1969),
         p. 110.

44.   Ibid., See also MEIKLEJOHN, D. "Liability for Oil Pollution
         Cleanup and the Water Quality Improvement Act of
         1970," Cornell Law Review, No. 55 (July 1970),
         p. 974.

45.   See U.S. vs. Republic Steel Corp. 362 U.S. 482, 80 S. Ct. 884
        and U.S. vs. Standard Oil Company 384 U.S. 224,
        86 S. Ct. 1427 as quoted in EDWARDS, op. cit., p. 111.

46.   See U.S. vs. Perma Paving Co. 332 F. 2d 754 (2d Cir. 1964) as
        quoted in University of California, Davis, op. cit.,
        p. 178.

47.   TRIPP, J. T. B. and HALL, R. M. "Federal Enforcement Under
        the Refuse Act of 1899," Albany Law Review, No. 35
        (Fall 1970), p. 80.

48.   HOVANESIAN, A. "Post Torrey Canyon: Toward a New Solution
        to the Problem of Traumatic Oil Spillage," Connecticut
        Law Review, No. 2 (Spring 1970), p. 635.

49.   HEALY, N. J. and PAULSEN, G. W. "Marine Oil Pollution
        and the Water Quality Improvement Act of 1970,"
        Journal of Maritime Law and Commerce, No. 1 (July
        1970), p. 538.

50.   MEIKLEJOHN, op. cit., p. 973.

51.   EDWARDS, op. cit., pp. 109-110.

52.   MEIKLEJOHN, op. cit., p. 973; O'CONNELL, op. cit., p. 121.

53.   University of California, Davis, op. cit., p. 177.

54.   HEALY and PAULSEN, op. cit., pp. 541-542.

55.   EDWARDS, op. cit., p. 109.

56.   MEIKLEJOHN, op. cit., p. 974.

57.   3 U.S.T. 2989, T.I.A.S. No. 4900, 327 U.N.T.S. 3 came into
        force July 26, 1958; United States Ratification, 1961.

58.   CLINGAN, T. A., Jr. and SPRINGER, R. "International Regula-
        tion of Oil Pollution," working paper for the Internation-
        al Law Panel of the President's Commission on Marine
        Science, Engineering, and Resources (undated), p. 23.

59.   "Oil Pollution of the Sea," Harvard International Law Journal,
        op. cit., pp. 341-342.

60. MEIKLEJOHN, op. cit., pp. 975-976.

61. 2 U.S.T. 1523, T.I.A.S. No. 6109 came into force May 18, 1967; United States Ratification, 1966.

62. HEALY and PAULSEN, op. cit., pp. 539-540.

63. O'CONNELL, op. cit., p. 123.

64. HOVANESIAN, op. cit., p. 637.

65. HEALY and PAULSEN, op. cit., pp. 543-544.

66. MEIKLEJOHN, op. cit., pp. 978-979.

67. HEALY and PAULSEN, op. cit., p. 544.

68. MEIKLEJOHN, op. cit., pp. 976-977.

69. President Nixon's Message to Congress on Oil Pollution, May 20, 1970, Bureau of National Affairs, Environmental Reporter (1970), pp. 241-243.

70. MEIKLEJOHN, op. cit., pp. 979-980.

71. Ibid., p. 980.

72. Ibid., p. 982.

73. Ibid., pp. 984-987.

74. HEALY and PAULSEN, op. cit., p. 551.

75. For an interesting account of the threat of oil pollution to coastal areas of Maine, New Brunswick, and Nova Scotia see McDONALD, J. "Oil and the Environment: The View from Maine," Fortune, No. 83 (April 1971), pp. 84-89, 146-147, 150; and "Proposal to Protect Maine from the Oilbergs of the 1970's," University of Maine Law Review, No. 22 (1970), pp. 481-510.

76. CUNNINGHAM, P. "Comment on the Washington Oil Spill Act of 1970," manuscript copy, forthcoming, University of Washington Law Review.

77.   State of Washington, Department of Ecology, Laws and Oil Spill
      Emergency Procedures, op. cit., pp. 2-3.

78.   Revised Code of Washington 90.48.340 (1970). Necessary ex-
      penses do not include expenses related to investigation
      or surveillance costs.

79.   Revised Code of Washington 90.43.135 (1970) and Conservation
      Foundation. Conservation Newsletter (November 1970),
      p. 11.

80.   CUNNINGHAM, op. cit.

81.   State of Washington, Department of Ecology, Laws and Oil Spill
      Emergency Procedures, op. cit., pp. 2-3.

82.   CUNNINGHAM, op. cit.

83.   Ibid.

84.   RODGERS, W. H., Jr. "Open Letter to the Legislature"
      (January 7, 1971), pp. 2, 4, 5.

85.   State of Washington, Legislature, 2nd Extraordinary Session, 1971.
      Coastal Waters Protection Act of 1971, Chapter 180,
      Laws of Washington.

86.   State of Maine. An Act Relating to Coastal Conveyance of Pet-
      roleum, H.P. 1459, L.D. 1835 (February 5, 1970);
      State of Maine. An Act to Regulate Site Location of
      Developments Substantially Affecting Environment,
      H.P. 1458, L.D. 1834 (February 5, 1970). These laws
      are quoted from HARRIS, M. B. and LOVETT, A.
      "Recent Developments in the Law of the Sea: A Syn-
      opsis," San Diego Law Review, No. 7 (July 1970),
      p. 631.

87.   Ibid.

88.   Ibid.

89.   State of Michigan. Annual Statutes, 14. 258, Sec. 2, Sec. 3
      (1970).

90.   Ibid.

91.   CUNNINGHAM, op. cit.

92.   United States, Department of Transportation, United States Coast
       Guard, 13th Coast Guard District. Seattle Coastal
       Region Oil and Hazardous Materials Pollution Contin-
       gency Plan. Seattle (December 1, 1970).

93.   United States, Department of Transportation, United States Coast
       Guard. Maritime Environmental Protection Activities
       of the Coast Guard, Commandant Notice 3010 (August 20,
       1970).

94.   State of Washington, Department of Ecology. Oil Spill Action
       Plan. Olympia: State of Washington (1972).

95.   Personal communication, Lieutenant James Phaups, Oil Pollution
       Officer, 13th Coast Guard District (April 15, 1971).
       Industry has begun planning to contain oil spills by
       accumulating equipment at Seattle, Tacoma, Everett,
       Anacortes, and Bellingham. See PAGE, D. "Oil
       Companies Unite to Contain Possible Spills," Seattle
       Post-Intelligencer (April 17, 1971), p. 3.

96.   United States, Department of Transportation (August 20, 1970),
       op. cit.; PAGE, D. "Puget Sound Marine Traffic
       Control Due," Seattle Post-Intelligencer, September 22,
       1971, p. A5.

97.   LEDERMAN, W. R. "The British North America Act, 1867, as
       amended, Sections 91-95," in LEDERMAN, W. R.
       (ed.) The Courts and the Canadian Constitution.
       Toronto: McClelland and Stewart, Carleton Library
       Series, No. 16 (1964), pp. 13-18. The role of the
       federal government has been analyzed by J. W. Mac-
       Neill. He concludes that a federal response will be
       necessary in almost all cases of environmental degrada-
       tion, especially those involving international compli-
       cations or conflicts between provinces and federal re-
       sponsibilities within a province. See MACNEILL, J.W.
       Environmental Management. Ottawa: Information
       Canada (1971).

98.   SMITH, H. A. Federalism in North America. Boston: Chipman
       Law Publishing Company (1923), pp. 9-13.

99. The problem of offshore mineral development and the related prob-
     lems of inland waters, territorial waters, and the con-
     tinental shelf are analyzed well by LaForest. See
     LaFOREST, G. V. Natural Resources and Public Pro-
     perty under the Canadian Constitution. Toronto:
     University of Toronto Press (1969), pp. 85-107.

100. Personal communication, Timothy O'Riordan, Department of Geog-
      raphy, Simon Fraser University (June 1, 1970). See
      also LEDERMAN, W. R. "Factors Conditioning Growth:
      Administrative and Jurisdictional Factors," in Resources
      for Tomorrow Conference. Proceedings, Vol. 3.
      Ottawa: Queen's Printer (1962), pp. 41-43; and
      comments by Pierre-Elliot Trudeau and Hugh J. Whalen
      at pp. 43-46.

101. SEWELL, W. R. D. "Multiple-Purpose Development of Canada's
      Water Resources," Water Power, No. 14 (April 1962),
      p. 146.

102. DUNN, J. D. "Oil Pollution of the Seas," Oil and Gas Seminar
      Paper. Vancouver: University of British Columbia,
      Faculty of Law (April 30, 1970), pp. 33-34.

103. Personal communication, the Hon. Ray Williston, Minister of
      Lands, Forests, and Water Resources for British Colum-
      bia (April 1, 1971), and the reply from the Minister
      (April 16, 1971).

104. ESSO Petroleum Company Ltd. vs. Southport Corporation, 1956
      Admiralty Court 218, as quoted in DUNN, op. cit.,
      pp. 8-11.

105. Supreme Court of Canada, City of Portage La Prairie vs. B.C.
      Pea Growers Ltd., 1966 Dominion Law Reports, 2nd,
      503, S.C.R. 150, as quoted in DUNN, op. cit.,
      p. 12.

106. Ibid., p. 12. Dunn's paper is one of the few papers analyzing
      oil pollution law in Canada.

107. Ibid., pp. 17-18. See also Canadian Council of Resource Min-
      isters. A Digest of Environmental Pollution Legislation
      in Canada. Montreal (May 1970), 2 vols., Report
      prepared for and under the direction and responsibility

of Canadian Industries Limited.

108.   Canada, Parliament. Statutes, 1868, chapter 60, section 14.

109.   Canada, Parliament. Statutes, 1886, chapter 95; 1894, chap-
        ter 51; 1895, chapter 27; 1906, chapter 45, Revised
        Statutes; 1914, chapter 8; 1927, chapter 27, Revised
        Statutes; 1932, chapter 42; 1952, chapter 119, Revised
        Statutes.

110.   Canada, Parliament. Statutes, 1960-61, chapter 23.

111.   Canada, Parliament. Statutes, 1969-70, chapter 63, section 3
        (1).

112.   Ibid., chapter 63, section 3 (8), (9), (10).

113.   Canada, Parliament. Statutes, 1917, chapter 18, section 1.

114.   Canada, Parliament. Statutes, 1921, chapter 39, section 1.

115.   Canada, Order in Council, August 3, 1966, No. 1475. The or-
        iginal regulation was introduced as Canada, Order in
        Council, August 17, 1948, No. 3632.

116.   "The Participation of the Government of Canada in the Investiga-
        tion and Abatement of Water Pollution," Background
        Papers Prepared for the Conference on Pollution and
        Our Environment," Montreal, October 31-November 4,
        1966. Montreal: Canadian Council of Resource Min-
        isters (1966), Vol. 2, paper B5-1, p. 12.

117.   Canada, Parliament. Statutes, 1968-69, chapter 15, section 10.

118.   Canada, Parliament. Statutes, 1873, chapter 65; 1874, chapter 29;
        1880, chapter 30; 1886, chapter 36; 1899, chapter 31;
        1906, chapter 115, Revised Statutes; 1927, chapter 40,
        Revised Statutes; 1952, chapter 193, Revised Statutes.

119.   Canada, Parliament. Statutes, 1968-69, chapter 15, section 14.

120.   DUNN, op. cit., p. 20.

121.   Canada, Parliament. Statutes, 1969-70, chapter 52, section 8.

122. Ibid., chapter 52, section 2.

123. Ibid., chapter 52, sections 22 and 23.

124. DUNN, op. cit., p. 22.

125. Canada, Parliament. Statutes, 1934, chapter 44.

126. Canada, Parliament. Statutes, 1956, chapter 34, sections 25 and 29.

127. Canada, Parliament. Statutes, 1964-65, chapter 39, section 31.

128. Canada, Parliament. Statutes, 1968-69, chapter 53, section 24.

129. "The Participation of the Government of Canada in the Investigation and Abatement of Water Pollution," op. cit., pp. 13-14.

130. British Columbia. Sanitary Regulations Issued under the Health Act, British Columbia Regulations, 142/59, section 66.

131. British Columbia, Legislative Assembly. Revised Statutes of British Columbia, 1960, chapter 405, section 41.

132. LUCAS, A. R. "Water Pollution Control Law in British Columbia," University of British Columbia Law Review, No. 4 (May 1969), p. 64. Lucas' article is a good summary of pollution control in British Columbia.

133. British Columbia, Legislative Assembly. Statutes, 1956, chapter 36.

134. Ibid., chapter 36, section 2.

135. LUCAS, op. cit., p. 66.

136. British Columbia, Legislative Assembly. Statutes, 1965, chapter 37, sections 2 and 7.

137. British Columbia, Legislative Assembly. Statutes, 1967, chapter 34.

138. Ibid., chapter 34, section 2.

139.   In 1970 the definition was further broadened but again no specific
            reference was made to oil pollution. See British Colum-
            bia, Legislative Assembly. Statutes, 1970, chapter 36,
            section 1.

140.   British Columbia, Legislative Assembly. Statutes, 1968, chap-
            ter 38, section 8.

141.   Personal communication, the Hon. Ray Williston, Minister of
            Lands, Forests, and water Resources for British Colum-
            bia (April 16, 1971).

142.   Canada, Parliament. Statutes, 1969-70, chapter 47.

143.   KONAN, R. W. "The Manhattan's Arctic Conquest and Canada's
            Response in Legal Diplomacy," Cornell International
            Law Journal, No. 3 (Spring 1970), p. 190.

144.   Statement by Prime Minister Trudeau. See Canada, Parliament,
            House of Commons. Official Report of Debates,
            vol. 6, 2nd session, 28th Parliament (April 8, 1970),
            p. 5624. The Act is discussed in detail by Daniel
            Wilkes. See WILKES, D. "International Administra-
            tive Due Process and Control of Pollution - The Cana-
            dian Arctic Waters Example," Journal of Maritime
            Law and Commerce, No. 2 (April 1971), pp. 499-539.

145.   Canada, Parliament. Statutes, 1970-71, chapter 27.

146.   Canada, Department of Transport, Marine Operations. "Interim
            Federal Contingency Plan for Combatting Oil and
            Toxic Material Spills." Ottawa: Department of Trans-
            port (1970).

147.   Ibid., p. 4.

148.   Ibid., p. 5.

149.   GOLDIE, L. F. E. "Amenities Rights - Parallels to Pollution
            Taxes," Natural Resources Journal, No. 11 (April
            1971), pp. 274-280.

150.   O'CONNELL, op. cit., p. 126. For a discussion of unilateral versus
            international action on the oil pollution problem see
            NEUMAN, R. H. "Oil on Troubled Waters: The International

Control of Marine Pollution," Journal of Maritime
Law and Commerce, No. 2 (January 1971), pp. 349-361.

151. HOVANESIAN, op. cit., p. 635.

152. United Nations Press Release/L.T./587/Rev, 1, April 9, 1970 as
      quoted in NEUMAN, op. cit., p. 356.

153. Canada, Parliament, House of Commons, Special Committee on
      Environmental Pollution. Minutes of Proceedings and
      Evidence. Ottawa: Queen's Printer and Information
      Canada (1970), Issues 2-13.

CHAPTER 4
INTERNATIONAL MECHANISMS FOR THE PREVENTION
AND CONTROL OF OIL POLLUTION

National regulations and standards are but one approach to the problem of preventing and controlling oil pollution and compensating victims of spills. There are also a number of international organizations, treaties, conventions, and authorities (in addition to the 1954 and 1962 Conventions discussed in the previous chapter) that address themselves to oil pollution. These institutions are important in the study area for several reasons. First, oil spilled on the high seas outside of the authority of either Canada or the United States could spread into the study area. In such an instance, the only sanctions which could presently be applied to the polluter are those agreed upon by the international community. Second, international organizations, both private and governmental, might prove helpful in partially assuming damage costs, thereby removing some of the burden suffered by individuals from spilled oil. Third, international agreements have the advantage of wider, though not necessarily global, applicability to the externality problem created by oil spills on the high seas or in territorial waters.

The first part of this chapter discusses the various international organizations and treaties relating to oil pollution. These include the Comite Maritime International (CMI), the Intergovernmental Maritime Consultative Organization (IMCO), the 1954, 1962, and 1969 Conventions, the applicable Geneva Conventions of 1958, and the Tanker Owners Voluntary Agreement Concerning Liability for Oil Pollution (TOVALOP). The desirability of international agreement is noted, and attention is given to the multitude of factors which retard multilateral agreements to prevent and control oil pollution. These factors were outlined in Chapter One and are given greater articulation in articles by Livingston, Wolman,

Schachter and Serwer, and Ross.[1] Specific complications, seemingly bla-

tantly apparent in oil transportation, can also arise. This is well docu-

mented in the Torrey Canyon case where the wrecked tanker:[2]

> ... carrying a cargo of 117,000 tons of Kuwait crude
> oil, was American owned and chartered, Liberian
> registered, manned by an Italian master and crew,
> contracted for salvage to a Dutch company, grounded
> on the Seven Stones reef in international waters,
> abandoned by the owners and destroyed by the
> British naval and air force using rockets and napalm.

In addition, coastal damage was incurred by both Britain and France. The

number of nations involved make settlement of pollution claims difficult

in the absence of stringent international controls. Treaties and organiza-

tions have also demonstrated a marked bias in favor of shipping interests

and freedom of the seas. Nations with long coastlines do not enjoy the

same protection afforded users of the high seas, and are forced to internal-

ize damage costs from spilled oil, costs which emanate from areas outside

that nation's control.[3] As a result, multilateral treaties on an internation-

al global scale have not resulted in equitable and fair compensation for

oil spill damage.

The second part of this chapter focuses attention on the develop-

ment of bilateral and regional arrangements as a more practical approach

to the prevention and control of marine pollution, especially in the study

area. The regional approach to pollution problems has long been impor-

tant in the control of river pollution for example.[4] International commis-

sions designed to protect international waterways have formed the basis for

existing cooperation. The International Commission for the Protection of

the Rhine Against Pollution, the International Joint Commission, and the

International Boundary and Water Commission are three examples of such

cooperation. Commissions which involve only two countries have been

more successful in gaining political approval of their recommendations.

For example, a report of the International Joint Commission for the control of pollution in the Great Lakes was accepted by both Canada and the United States, and initial steps have been taken to implement the report.[5] A similar agreement was signed by the United States and Mexico when the quality of water delivered to Mexico from the Colorado River deteriorated rapidly. Saline waters from the Colorado basin could not be used in Mexican irrigation works. To solve the problem, the United States agreed, at its own expense, to build a bypass around the Mexican diversion works. Multilateral commissions have made studies of water quality, but many lack authority to determine who is responsible and who will pay for pollution control. The International Commission for the Protection of the Rhine Against Pollution was formed in 1950 by Switzerland, France, Luxembourg, the Federal Republic of Germany, and the Netherlands, but was limited to the joint testing of water quality. The Commission did little to control pollution, and increasing industrial development on the lower Rhine only intensified the problem.[6]

Advocating a regional approach to marine pollution is not new. In 1968 Jackson argued that a series of regional sea-area agreements might turn "out to be the most effective approach to deliberate discharge of pollutants." He specifically excluded oil and radioactive wastes because of the global nature of these problems, but noted that regional agreements might usefully "be made within the context of a general global agreement or at least a declaration of principle."[7] Busch and Mears outline the basic advantages inherent in a regional approach. First, bilateral or multilateral agreements between industrialized nations would be of help in providing guidelines for developing nations and future international conventions. Second, the regional approach would meet the immediate pollution problems of countries sharing common interests in waters in one geographic area. Third, regional arrangements would avoid extensive debate over provisions dealing with waters that are miles distant. Fourth, such arrangements would facilitate more rapid actions against pollution.[8] Schachter and Serwer argue that "marine pollution is, of course, but a part of the totality of environmental problems that confront us today. A purely piecemeal approach, characterized by approaching a single problem, without considering its relationship to others, would not be adequate." Given these constraints, we should also recognize the "need for regional pollution control organs since it is apparent that, although pollution is a global problem, it is not uniformly global."[9] In spite of these recommendations, few concrete attempts have been made to manage pollution on either a bilateral or multilateral scale. The few attempts that have been made concentrate on a limited number of pollutants such as oil. By focusing on one pollutant (as is attempted in the Canada Shipping Act) it may then be possible to establish protective measures, enforcement mechanisms, and compensation procedures which could later be applied to a broader list of pollutants.

This chapter argues that little can or has been accomplished to control oil pollution on a global scale, given the polarization of nation-states between interests seeking to protect the freedom of the seas and those desiring to protect the marine environment of coastal states, outside the enunciation of general principles. It also contends that the most productive efforts of oil pollution control have come (and are likely to continue to emanate) from domestic laws and regional treaties. The intensification of efforts in these two areas will, it is argued, lead to more rapid agreement on a global scale.

MULTILATERAL APPROACHES TO OIL POLLUTION PREVENTION AND CONTROL

Almost all sources of international law lend support to the tenet that injuries to the beneficial uses of water in another state are not permissible.[10] This concept has its roots in international river pollution[11] but it is now regarded to be more generally applicable to all water pollution.

158

With such a concept, the key to international pollution control is that there is only a breach of international law when there is injury to the beneficial uses of the sea. Thus pollution of the seas by vessels of another state would not be a breach of international law unless some other use were injured. In addition, jurisdiction over activities on the high seas and prosecution is vested in the state of the flag. This raises two distinct problems. Flag states appreciate the relaxed shipping rules and are reluctant to rigidly enforce pollution violations, especially if such enforcement would threaten revenue derived from shipping sources. Other states, threatened by oil spills from outside territorial waters and unable to rely on flag states for adequate prosecution, might resort to extreme measures of self help.[12] Legault has characterized this problem as a conflict between "flag states" and "coastal states." Flag state interests:

> are preponderantly global rather than coastal. The system of law they have developed seeks to restrict the assertion of coastal authority, both quantitatively and qualitatively, in order to maintain an essentially laissez-faire regime conducive to the operation of their great navies and merchant fishing fleets.

Coastal states have, on the other hand, "their maritime interests primarily centered on their own coasts" and have "in varying degrees sought to extend the scope of coastal authority."[13] Maritime pollution, while a matter of universal concern, is of greater interest to coastal states. Such a classification is arbitrary and is neither absolute nor all-embracing, but the conflict between states is nonetheless real and is reflected in the international organizations and treaties governing maritime affairs. Moreover, it has helped frustrate attempts to establish a permanent international maritime organization with supervision over all aspects of international maritime trade and transport.[14]

International Organizations

The oldest maritime organization still in existence is the Comité

Maritime International. Originally established in 1897 and headquartered in Brussels, it attempts to promote the unification of international maritime and commercial law and practise, either through treaty or convention or by establishing uniformity of domestic laws, usages, and customs. It is a nongovernmental body composed of maritime law associations from about 30 countries, the majority of which are economically developed.[15] The law associations help draft conventions and then request the Belgian government to convene a diplomatic conference. If the proposed convention is adopted by the conference, national governments are expected to ratify the convention and introduce domestic legislation to incorporate principles of the draft convention. Major problems arose, however, because CMI conventions were drafted by shipowner representatives, and national governments or other interested parties were seldom privy to the convention during drafting. As a result, many of the conventions failed to achieve ratification by many of the major maritime nations, save those who are primarily "flag states."[16] Of special concern here is the failure of CMI to consider oil pollution a major international marine problem. At best it would seem that CMI could organize shipowners to police their own preventive and compensatory scheme, for without broader representations in drafting conventions, national governments are not likely to treat CMI deliberations and conventions very seriously.

The other major maritime organization, the Intergovernmental Maritime Consultative Organization, has taken more positive measures to prevent oil pollution. IMCO is a specialized organization, under the auspices of the United Nations, concerned solely with maritime affairs. It was established in 1958 with headquarters in London. The purpose of IMCO is "to achieve the highest practicable standards of maritime safety and efficient navigation by facilitating cooperation among governments in technical matters of all kinds affecting shipping" and "to discourage discriminatory, unfair and restrictive practises affecting ships in international

trade so as to promote the freest possible availability of shipping services
to meet the needs of the world for overseas transport."[17] Among its other
functions, IMCO is also responsible for administering the 1954 and 1962
Conventions discussed earlier in Chapter Three. While it is a body with
only advisory and consultative power, there is a permanent assembly,
council, maritime safety committee, and secretariat working on pollution
problems. The organization was instrumental in drafting the 1969 International Convention Relating to Intervention on the High Seas in Cases of
Oil Pollution Casualties (hereafter cited as the 1969 Public Law Convention) and the 1969 International Convention on Civil Liability for Oil Pollution Damage (hereafter cited as the 1969 Civil Liability Convention).[18]
IMCO has not, however, addressed itself to all sources of marine oil pollution. Operations on the continental shelf have been largely ignored
while IMCO attempts to reduce accidental and deliberate oil discharge
from ships, tankers, and barges.[19] The organization lacks any enforcement powers over the Conventions and their administration. Disputes over
interpretation or application of the Conventions can be referred to the International Court of Justice in absence of settlement by negotiation or arbitration,[20] but, as noted earlier, prosecutions for violations of all oil
pollution conventions rest with the state of flag. In spite of new conventions agreed to in 1969 (and discussed later in this chapter), IMCO remains
powerless to enforce convention regulations and has no independent resources
to compensate oil spillage victims.

International Conventions

In addition to these organizations there are a number of international conventions which address themselves to marine pollution. The Geneva
Convention on the Continental Shelf of 1958 permits states to exercise sovereign rights over adjacent continental shelves for the purpose of exploring
and exploiting any possible natural resources. Exploration and exploitation of shelf resources must not result in any justifiable "interference with

navigation, fishing or with the conservation of the living resources of the sea" and further, the coastal state is obligated to undertake "all appropriate measures for the protection of the living resources of the sea from harmful agents" in safety zones extending up to 500 meters from installations used in the exploration and exploitation of the shelf.[21] Presumably, oil pollution caused by such exploitation of underground oil resources would constitute interference and be a violation of the Convention. There are no guarantees for prosecuting oil pollution violations under the Convention. Additional problems arise from the increasing ability of the developed nations to explore and exploit the seabed to limits beyond those established in the Convention.[22] Moreover, the increasing move seaward raises the possibility that two states may legitimately claim rights to the same offshore seabed.

Article 24 of the Geneva Convention on the High Seas requires states to "draw up regulations to prevent pollution of the seas by the discharge of oil from ships or pipelines or resulting from the exploration and exploitation of the seabed and its subsoil, taking into account of existing treaty provisions."[23] This provision was drafted with the 1954 Convention specifically in mind but states are under no compulsion to adhere to that convention or to enact similar legislation. The Convention represents, however, a codification of international law, and imposes a duty upon the contracting parties to take action to prevent oil pollution even if they are not party to the 1954 and 1962 Conventions.[24]

The third major Geneva Convention of 1958, the Convention on the Territorial Sea and the Contiguous Zone, gives the coastal state, in a zone of the high seas contiguous to its territorial seas, power to exercise the control necessary to prevent "enfringement of its ... sanitary regulations within its territory or territorial sea and punish enfringements of the above regulations committed within its territory or territorial sea." Restrictions are placed on the breadth of the contiguous zone. It may not extend beyond twelve miles from the baseline from which the breadth of the territorial sea is measured.[25] Thus foreign flag tankers spilling oil outside of twelve miles would not be subject to the Territorial Sea and Contiguous Zone Convention.

The 1954 and 1962 Conventions are more stringent in prohibiting oil pollution than the Geneva conventions, but not all nations are party to these oil conventions. Even those nations that are signatories have no guarantees of adequate prosecution of violations since this is left to the flag state. While offering greater universality and codifying international law, the three Geneva conventions discussed above do little to help control oil pollution. They express the responsibilities of states and their nationals to desist from polluting but there are no enforcement and policing procedures to prohibit oil spillage. One of the countries of concern in this study, Canada, while party to the oil pollution conventions of 1954 and 1962, has not ratified the Geneva conventions. Government spokesmen have expressed the view that Canada accepts the Geneva conventions as representing the current state of international law. In view of these statements and ratification of the 1954 and 1962 conventions Canada's refusal to ratify the Geneva conventions is of little consequence in reducing marine oil pollution. The United States, on the other hand, has ratified both the oil pollution and Geneva conventions.

Other conventions are more specifically concerned with oil pollution. These include the 1954 and 1962 Conventions, discussed previously in Chapter Three and incorporated in domestic legislation of both Canada and the United States. On November 13, 1969, forty-nine nations began a two week conference in Brussels, under the auspices of IMCO, to consider draft conventions concerning marine pollution. Two new conventions emerged from this conference; the 1969 Public Law Convention and the 1969 Civil Liability Convention.

The 1969 Brussels Conventions

The 1969 Public Law Convention[26] states that parties to the Convention may take such measures on the high seas as may be necessary to
prevent, mitigate, or eliminate grave and imminent danger to their coast
line or related interests from pollution or a threat of pollution of the sea
by oil, following upon a maritime casualty (or acts related to such a casualty) which may very reasonably be expected to result in major harmful
consequences. The actions of coastal states are restricted, however, by
procedural rules and sanctions for excessive conduct. No action can be
taken against warships or vessels owned and operated by a state in non-
commercial service. Before actions can be taken against other ships,
coastal states are required: to consult with other states affected by the
casualty, especially the flag state; to give notice of proposed measures to
other interested parties; and to consult with independent experts, a list of
whose names are to be compiled by IMCO. Notwithstanding these provisions, a state may, in an extreme emergency, take measures that are dic-
tated by the urgency of the situation without notification or consultation.
Even under these emergency provisions, the coastal state must notify interested nations, avoid risk of human life, and abstain from interference with
repatriation of the crew. Additional self-restraint is imposed by a standard for judging protective measures. Actions taken by the coastal state
must be proportionate to the damage actual or threatened. Any overaction
on the part of the coastal state will necessitate that state paying compensation for the amount of overaction. Controversies over application of the
Convention are to be settled through procedures of negotiation between
the parties, conciliation, and, if necessary, arbitration. The Convention
will enter into effect 90 days following the date that fifteen states have
signed it (without reservations as to ratification, acceptance, or approval)
and have deposited appropriate instruments with the Secretary General of
the United Nations.

The Convention is a compromise, but flag states relinquished little. The right of states to take measures to protect their shores from pollution "seems reasonably well grounded in current customary international law."[27] Inclusion of procedures for consulting other interested parties and resolving disputes serves as additional protection for the flag states. Minor concessions were granted to coastal state interests. States were given the right to include cases involving "major" damage rather than "catastrophic" or "disastrous" incidents only. Provisions restricting the rights of coastal states to take actions against pollution in their own territorial seas were defeated.[28] In addition, the Convention is limited in application to ships, (defined to include barges, seagoing vessels, and floating craft) but oil rigs conducting drilling operations are excluded from the terms of the Convention.[29]

The Convention was approved by a substantial majority of participating states; however, it is important to note in this study that Canada abstained while the United States voted in favor. While Canada may recognize the Convention (presuming it comes into force) as representing international law, as she has done with the Geneva conventions discussed above, her decision to withdraw the authority of the International Court of Justice for all disputes concerning marine pollution raises the possibility that she may not submit to binding arbitration, though she may still participate in voluntary negotiations should any dispute arise.

The 1969 Civil Liability Convention[30] attempts to establish uniform international rules and procedures for determining questions of liability and compensation. Application of the Convention is limited to pollution damage in the territory of a contracting state, including its territorial sea, caused by contamination resulting from the escape or discharge of persistent oil (such as crude oil, fuel oil, heavy diesel oil, lubricating oil, and whale oil) carried in a sea going vessel or any seaborne craft of any type whatsoever, actually carrying oil in bulk as cargo. Passenger vessels and dry cargo vessels are thus excluded, if they are not carrying oil as cargo. Warships and other state owned or operated ships used for non-commercial government service are excluded, but state owned ships in commercial service are subject to the Convention. Pollution damages include the costs of reasonable preventive measures taken by anyone after the incident to prevent or minimize such damage. Both government and private clean-up costs are thus included within the definition.

All liability is channelled through the registered owner of the ship at the time of the incident. Rights of recourse by the owner against third parties is preserved by the Convention. In the case of two ships colliding, the owners of both are jointly and severally liable for all pollution damage which cannot be reasonably separated. Liability is without regard to fault; however, it is "strict" rather than "absolute" in that the owner may be absolved under certain conditions. No liability results if the damage resulted from an act of war, a natural phenomenon of an exceptional, inevitable, and irrestible character, an intentional act or omission of a third party, or negligence or other wrongful act of an authority responsible for aids to navigation. The United States succeeded in opposing attempts to exempt liability in the case of negligent acts or omissions of third parties.

Liability for oil spillage under the Convention is limited. The owner of a ship shall be entitled to limit his liability under the Convention, in respect of any one incident, to an aggregate amount of approximately 135 dollars for each ton of the ship's tonnage. However, this aggregate amount shall not in any event exceed 14 million dollars.[31] If, however, the accident is the result of actual fault or privity on the part of the owner, there is no limit of liability. Owners may benefit from the limitation by establishing a fund (or an equivalent guarantee) in the courts of a contracting party, up to the applicable liability limit. Ships which carry over 2,000 tons of oil must maintain insurance or other financial security in an amount determined by applying the Convention's liability limits to the tonnage of the ship, and must carry a certificate attesting to such financial responsibility.

Claims for pollution damage may be pressed in the courts of any contracting party in whose territory or territorial waters damage has occurred or preventive measures have been taken. These courts are deemed competent by all contracting parties to award damages among all legitimate claimants. Any final judgement is enforceable in the courts of all the contracting states, and is not to be reopened except on a showing of fraud or denial of notice for an opportunity to be heard.

The Civil Liability Convention was approved by a vote of 34 to 1 with 10 abstentions. The United States voted for the Convention while Canada abstained.[32] The Convention will enter into force on the ninetieth day following the date on which eight states, including five states with not less than one million gross tons of tanker tonnage, have signed and ratified the Convention and deposited the appropriate instrument with the Secretary General of the United Nations.

The Civil Liability Convention is one of the few agreements to recognize the principle of extraterritorial damage and to take concrete steps to compensate victims. It does not, however, remove all of the externalities that can be created when oil spilled on the high seas causes damage to the territory of a coastal state. Claims may be made for damages, but the Convention is not specific when it comes to indirect damages. Cases would be tried in courts of the territory damaged, and the United States and Canadian courts have not been disposed to award indirect damages. Limits are placed on the amount of liability for the registered owner of the ship. Vessels carrying less than 2,000 tons of persistent oil in bulk are not covered. The Convention prohibits states from imposing their own standards of liability for spills occurring within territorial waters and provides an exclusive remedy for both government and private claimants, thus depriving private parties of their rights under general maritime law.[33] Moreover, the Convention has all the inherent limitations of unpoliced international agreement: states are under no compulsion to sign the Convention; there are no provisions for dealing with anonymous spills; and the Convention lacks specific preventive measures.

Many of the basic provisions of the Civil Liability Convention have already been enacted into domestic law in Canada and the United States through the Water Quality Improvement Act of 1970 and the Canada Shipping Act, 1971.[34] Liability limits, strict liability, requirements of evidence of financial responsibility, and payments for government cleanup costs are common features of both domestic laws and the Convention. Important differences can be noted, however, in the broadness of claims, compensation funds, claim procedures, policing measures, and persons liable.

While the Convention is not yet in force, and Canada and the United States have not ratified the Convention, it is interesting to speculate on the basis of each nation's vote at Brussels. If the United States ratifies the Convention, she would be able to collect damages in domestic courts for private interests as well as government costs. The Convention therefore protects a broader range of people than the Water Quality Improvement Act of 1970. If Canada does not ratify the Convention, she would have no recourse to claims for extraterritorial damage because she would not be a contracting party. Any ships from Alaska which might spill oil on the high seas and damage Canadian territory would be under no obligation to satisfy Canadian claims; however, if oil from the same incident damaged United States territory the vessel would be liable.

The 1971 Brussels Convention

The 1969 Brussels Conference, recognizing some of the inequities in the 1969 conventions, adopted a resolution requesting IMCO to convene

an international compensation fund. Representatives from over fifty countries met from November 29 to December 28, 1971 in Brussels and concluded a new convention entitled The International Convention on the
Establishment of an International Fund for Oil Pollution Damage,[35] which
is a supplement to the 1969 Civil Liability Convention. The fund is designed to ensure full compensation for all victims of oil pollution and to
remove some of the financial burden placed on shipowners under the Civil
Liability Convention.

The fund will compensate victims if persons have been unable to
obtain compensation under the Civil Liability Convention; however, it
will not accept absolute liability for all damage. Pollution damages from
an act of war, a government owned ship, or an intentional omission of a
third party are not covered by the fund. Unlike the Civil Liability Convention damage resulting from a natural phenomenon of an exceptional,
inevitable, or irrestible character is covered by the fund.

While the limits of compensation are increased through the fund
they are nevertheless limited. The aggregate amount of compensation for
any one incident is limited to 450 million francs or approximately 30 million dollars, although there is a mechanism by which compensation can be
increased to 900 million francs (60 million dollars). Amounts collected by
pollution victims under the Civil Liability Convention are to be included
in the 30 million dollar ceiling.

To relieve some of the financial burden placed on shipowners, the
Convention provides that the shipowner shall be indemnified for that portion of the aggregate amount of liability under the Civil Liability Convention which is in excess of 1,500 francs (100 dollars) for each ton of the
ship's tonnage or a total of 125 million francs (8,333,333 dollars), whichever is less, and is not in excess of 2,000 francs (133 dollars) for each ton
of the ship's tonnage or a total of 210 million francs (14 miliion dollars),
whichever is less.

Contributions to the fund are to be made by individuals receiving more than 150,000 tons of oil per year. Initial contributions would be limited to 75 million francs (5 million dollars). Annual contributions are to be set by a governing Assembly on a per ton basis.

The fund will remedy some of the deficiencies in the Civil Liability Convention but it will not ensure full and adequate compensation as intended. The Convention places an unrealistic limit on financial responsibility. Damage from a postulated supertanker mishap in the Strait of Georgia, for example, is estimated to cost Canada 100 million dollars.[36] This figure, while perhaps inflated, nevertheless does not include damage that would occur in the United States or aesthetic and wildlife losses. Even if damages were only half that claimed in the Canadian Aide-Memoire to the United States the fund established by the 1972 Convention would not be adequate to cover the damages.

Earlier, on October 21, 1969, the Assembly of IMCO convened in London and adopted amendments to the 1954 Convention. These amendments, which are not yet in force, prohibit discharges from ships beyond a negligible permissible amount in all areas of the sea. Tankers (meaning ships in which the greater part of the cargo space is constructed or adapted for the carriage of oil in bulk) are prohibited from discharging any amounts of oil within 50 miles of land. The discharge of oil or oily mixtures from machinery space bilges on tankers, however, is governed by the same rules that apply to ships. These amendments will come into force twelve months after they have been ratified by two-thirds of the governments that are party to the Convention.[37] The new provisions broaden the zone of prohibited discharge, but the basic criticisms noted in Chapter Three still apply: surveillance is weak; reporting is left to the master of the ship; and penalties are not specified and are left to the state of flag.

Private Control Programs

Beyond the international agreements, many of the corporations

engaged in the world trade of petroleum have instituted their own plans
to complement existing conventions. On January 7, 1969, the major oil
tanker owners announced signing of an insurance scheme designed to compensate governments for cleanup efforts and to encourage owners and operators to take remedial measures. Known as the Tanker Owners Voluntary
Agreement Concerning Liability for Oil Pollution Damage (or TOVALOP
and hereafter cited as such), the plan provides reimbursement to national
governments only if the tanker is operated negligently. Onus is placed on
the owners to prove that the tanker was not operated negligently. No
compensation is given to third parties or private individuals who incur
costs in cleanup, and compensation applies only to physical contamination.
Fire or explosion damage, consequential damage, or ecological damage,
is not covered under the scheme.[38] Reimbursement for costs is financed
through a Bermuda-based indemnity association with liability limits of
100 dollars per gross registered ton or 10 million dollars, whichever is
less.[39] From this total liability, the tanker owner is allowed to deduct
the expenses he incurs in helping to clean up and control the spilled oil.
TOVALOP is now administered by the International Tanker Owners Pollution Federation with headquarters in London. The Plan came into effect
when 50 percent of the tankers of the world (excluding tankers owned by
a government or government agency and tankers of under 5,000 deadweight tons) became parties.[40]

On January 14, 1971, the scheme was supplemented by a new
contract which took into account the Civil Liability Convention passed at
Brussels in November, 1969. Persons eligible for compensation were
broadened to include any individual or partnership of any public or private body whether corporate or not, including a state or any of its con-
stituent subdivisions. The compensation limit was revised, but, under the
contract, is not to exceed 30 million dollars less the sum of the following:
1.   the owner's maximum liability under TOVALOP,

2. the amount of expenditures for which the owner was entitled to re-
    ceive reimbursement for cleanup expenses as provided in TOVALOP,

3. the maximum liability for the owner with respect to such damage
    under applicable law, statutes, regulations, or conventions,

4. the maximum amount that persons sustaining pollution damage were
    entitled to receive from other sources.[41]

Compensation claims remain limited to physical contamination, and there
are no provisions for claims in excess of the liability limits. There is,
therefore, no guarantee, even with the Civil Liability Convention and
TOVALOP, that all externalities created by spilled oil will be internalized.

Regional Agreements

Some attempts have been made at the multilateral level to deal
with oil problems in specific critical areas. On June 6, 1969, Belgium,
Denmark, France, the Federal Republic of Germany, the Netherlands,
Norway, Sweden, and the United Kingdom signed the Agreement Concerning Pollution of the North Sea by Oil. The agreement generally provides for mutual aid and assistance in discovering oil pollution. Countries
have agreed to exchange information in dealing with oil pollution, and to
cooperate in its disposal. Sweden has also attempted to organize meetings
to discuss coastal oil pollution in the Baltic.[42] Attempts have also been
made to combat regional pollution in the Mediterranean and Persian Gulf
areas but these regional groupings are not as far advanced as international
river basin authorities. Data collection, program coordination, and mutual aid plans as provided for in these agreements are necessary prerequi-
sites to international control and prevention of oil pollution, but regional
pacts limited to such functions cannot achieve rapid success in reducing
oil pollution.

Multilateral attempts to prevent and control international oil pollution have not had any more success than the domestic responses discussed

earlier in Chapter Three. Like the domestic actions, international regulations focus on prohibiting discharges and establishing sanctions. Such regulations are far from perfect: they do not ensure full compensation or global coverage; there is a lack of preventive measures; the recognition of the public's right to a pollution free environment is nonexistent; and there are no contingency plans for dealing with spills on the high seas. Prosecution involves long and tedious procedures and rests with those states having the least interest in pollution control. Multilateral agreements such as TOVALOP and the Civil Liability Convention will complement domestic actions, but they will not internalize all the costs because of the limitations discussed above.

Most of the multilateral agreements focus on oil discharged on the high seas and subject to international law. The threat of this type of international oil spill exists in the study area, and will increase if the Alaska pipeline is built and if the bulk of the oil is destined for the United States west coast, but the greatest threat of transboundary oil pollution currently arises from vessels transporting oil in Puget Sound and the Strait of Georgia. Nations have the right, under the Geneva conventions of 1958, to develop sanitary regulations for their own territorial waters, and are under an obligation to see that their territory is not used as a source of extraterritorial damage. In the absence of comprehensive international law, the practical recourse to nations facing potential pollution problems is to explore the possibility of bilateral arrangements for prevention and control.

BILATERAL APPROACHES TO OIL POLLUTION CONTROL IN THE STUDY AREA

Bilateral agreements to manage common property resources have been most common in river basins, and the most successful treaties have had a limited number of participants. Relations between such countries have been cordial, and there exists a long history of consultation and

173

agreement on border problems. Canada and the United States participate in a number of joint pacts which regulate or seek to improve the management of common property resources. These include agreements relating to the International Joint Commission, the Pacific Salmon Commission, the Pacific Halibut Commission and the Migratory Birds Convention. There are, however, no international agreements regulating the quality of the marine environment or preventing and controlling oil pollution in the territorial waters of Canada and the United States. The increasing threat of oil pollution in the study area will provide a test of each nation's desire to reduce the potential for international oil pollution damage in territorial waters. If Canada and the United States (with their long history of cooperation on border problems) fail to deal with regional oil pollution problems, then prospects for comprehensive international controls at the regional and global scales would be bleak. If, on the other hand, a comprehensive plan can be achieved, application to other problem areas might serve to enhance efforts to achieve better international standards. This section of the chapter examines the appropriateness of existing agreements and institutions for any future program that seeks to manage the marine environment in Puget Sound and the Strait of Georgia.

Attention was given in Chapter One to the lack of concern with international pollution until the 1960's and the slow pace at which the international community recognizes global problems. This ignorance is reflected well in the report of Arnold Heeney and Livingston Merchant entitled Canada and the United States: Principles for Partnership. Commissioned to investigate "the desirability and practicability of developing acceptable principles which would make it easier to avoid divergencies in economic and other policies of interest to each other,"[43] the authors recommend policies for consultation and include a list of cases to be examined. Pollution problems along the boundary are manifold, yet the report did not consider the problem worthy of being studied as a case problem.

Water Pollution Along the Boundary

The general concern for transboundary pollution has not been man-
ifested in concrete plans for action until the 1970's. In the mid 1960's
the International Joint Commission initiated, at the request of the United
States and Canada, a study to investigate the pollution problem in the
Great Lakes. Several interim reports have been submitted to both governments documenting the rapid decline in the quality of water in the Great
Lakes and in the international section of the St. Lawrence River. The
third interim report, issued on May 21, 1970, dealt with the problem of
actual and potential oil pollution in these waters as a result of underwater
oil and gas drilling, and called upon the United States and Canada to ac-
celerate research programs and develop integrated contingency plans for
dealing with spills.[44] As a result of these reports, a meeting was held on
June 10, 1971 in Washington and both nations have now agreed on plans
for cleaning up the Great Lakes. In 1972 President Nixon and Prime Minister
Trudeau formally signed a Treaty implementing these plans. Officials have expressed the hope that the treaty will lay the basis for dealing with more
pollution problems along the boundary in the 1970's.[45]

Water pollution problems along the Canada-United States boundary
have traditionally come under the aegis of the Boundary Waters Treaty of
1909 and the International Joint Commission, but they are not capable of
dealing with marine oil pollution on a continuing basis because of restrictions in the treaty. New treaties are not expected to deal with matters of
legal jurisdiction in cases of extraterritorial damage, and no agreement is
expected to enforce pollution damage settlements won by an injured party
in one country against a polluter in the other.[46] Reports to the President
on oil spillage have advocated that controls over the discharge of oil and
other hazardous substances into international boundary waters of the United
States and Canada be reviewed and strengthened. These reports also suggest international agreements be consummated among the United States,

Canada, and Mexico for the control of pollution incidents; but these aspects of the reports have been largely ignored.[47] Without radical changes in government policies, little action can be taken in the 1970's to reduce marine pollution through existing treaties and institutions.

Transboundary Pollution and the International Joint Commission

The purpose of the Boundary Waters Treaty of 1909 between the United States and Canada is to:[48]

> prevent disputes regarding the use of boundary waters and to settle all questions which are pending between the United States and the Dominion of Canada involving the rights, obligations or interests of either in relation to the other or to the inhabitants of the other, along their common frontier, and to make provisions for the adjustment and settlement of all such questions as may hereafter arise.

To facilitate agreements, the Treaty established the International Joint Commission which has jurisdiction over and can pass on all cases involving the use, obstruction, or diversion of the United States-Canadian boundary waters, including water pollution.[49] The Treaty itself provides that all problems investigated by the Commission must first be referred to it by both governments. In all of these references, however, "boundary waters" has been limited to fresh waters along the course of which, or through which the international boundary runs. The International Joint Commission has yet to study and report its findings on any marine pollution problem.[50]

This precedent should not be construed as binding the International Joint Commission to only fresh water boundary problems. Article IX of the Boundary Waters Treaty authorizes the Commission to investigate and report upon particular questions and matters referred to it, with appropriate conclusions and recommendations.[51] The authority of the Commission is therefore restricted to matters referred to it by both governments, and the recommendations are not in themselves binding upon either of the

governments. The Commission may, depending on the type of case referred
to it, perform judicial, investigative, administrative, and arbitrative
functions.[52] Although questions may be referred to the Commission for
decision or arbitration, pursuant to Article X, neither government has
made use of this Article.[53]

Under Article IX, governments have referred water pollution problems in lakes Erie and Ontario, the connecting channels of the Great
Lakes system, and the Saint Croix, Rainy, and Red rivers. Boundary air
pollution problems near Trail and Detroit have also received attention
from the International Joint Commission.[54] References to water pollution
in the Treaty would not usually apply to the pollution of coastal waters
by oil; however, the language used in Article IX has been used to justify
a role for the International Joint Commission in many types of boundary
problems.[55] At present the Commission has no authority to investigate oil
pollution in coastal waters near the boundary without a specific reference
from both governments. Even if it were given power to investigate, recommendations would not be binding, and the Commission would have no policing powers without a specific reference. It is apparent, therefore, that
if there is to be any bilateral body to regulate and control pollution along
the border, "there will have to be a neoteric and fundamentally different
treaty agreed upon by the two countries."[56]

Such a treaty will have to reflect a marked change in the attitudes
and political desires of Canada and the United States to solve their common
pollution problems. Environmental questions have only been referred to
the Commission when the problems reached a crisis and were beyond the
capabilities of national and subnational governments. Emphasis, therefore, should be placed not on a lack of agreement with respect to oil pollution or other environmental problems, but on the lack of political will
and the reluctance of these nations to surrender domestic powers to an
international organization. When the Commission has been given power

to investigate problems, it has clearly demonstrated that thorough research is fundamental to any management program. The emphasis on research and the experience in dealing with boundary problems has given the Commission the ability to anticipate problems, but, under the present Treaty, the Commission is impotent without the consent of national governments. Moreover, the record of the Commission in adjudicating boundary problems fairly has given it a measure of respect in both countries far beyond its limited powers. Commissioners have acted outside of purely national constraints and have viewed problems from the international perspective. Any new arrangements or institutions to manage transboundary resources on a continuing basis would do well to emulate these characteristics of the International Joint Commission.

SUMMARY

Efforts of multilateral organizations remain focused on partial attempts at compensation after spillage. Some agencies, such as IMCO, have made attempts at preventing and controlling potential oil spills, but international agreement and enforcement is less well developed in this area than in compensating victims of oil spillage. International agreement on prevention, control, and compensation is desirable, but little has been accomplished to meet the problems posed by oil discharge on the high seas. Domestic compensation laws, contingency plans, and preventive measures are currently far more advanced and better prepared to meet spillage problems. International agreements are only as stringent as those seeking to protect the freedom of the seas will permit. The political constituency seeking to protect the global environment, particularly the interest of coastal areas, is comparatively less well organized. This constituency has not been able to pressure national governments into firm action in prevention and control, and has only partially succeeded in achieving compensation for spills. International compensation funds, as provided in the Civil Liability Convention and TOVALOP, will help to internalize some of the externalities. However, these funds will not alleviate the problem, since in most instances they parallel the most restrictive features of domestic legislation in Canada and the United States: liability is limited; enforcement procedures are weak; indirect damage is not covered; and fault must usually be proven.

Regional approaches to pollution control have been undertaken in other areas of the world with varying degrees of success. There is an absence of an effective regional arrangement to control marine oil pollution beyond data exchange centers. In the Puget Sound and Strait of Georgia area, however, international mechanisms for consultation and implementation do exist, and with only two countries involved, these institutions may help pave the way to some localized international agreement on oil pollution. It is apparent that there is a need for a new type of international agreement to deal comprehensively with oil pollution; not an agreement which will retard progress to more effective global controls, but one that will meet current needs more effectively. Chapter Five considers the alternatives to existing institutions and programs at the domestic and international levels.

REFERENCES

1. LIVINGSTON, D. "Pollution Control: An International Perspective," Scientist and Citizen, No. 10 (September 1968), pp. 172-182; WOLMAN, A. "Pollution as an International Issue," Foreign Affairs, No. 47 (October 1968), pp. 164-175; SCHACHTER, O. and, SERWER, D. "Maritime Pollution Problems and Remedies," American Journal of International Law, No. 65 (January 1971), pp. 84-111; ROSS, W. M. "The Management of International Common Property Resources," Geographical Review, No. 64 (July 1971), pp. 325-338.

2. NANDA, V. P. "The Torrey Canyon Disaster: Some Legal Aspects," Denver Law Journal, No. 44 (Summer 1967), p. 401.

3. LEGAULT, L. H. J. "The Freedom of the Seas: A License to Pollute?," paper presented at the Symposium on International Legal Problems of Pollution under the auspices of the Canadian Branch of the International Law Association, the University of British Columbia, and the Department of External Affairs, Ottawa and held in Vancouver, September 8-11, 1970, pp. 1-3.

4. LANGRAN, K. J. The Political and Administrative Control of Water Pollution in International River Basins. Seattle: University of Washington, Department of Geography, unpublished M.A. Thesis (1968).

5. ABLETT, D. "U.S., Canada Nearing Pact on Cleaning Up Great Lakes," Vancouver Sun, June 9, 1971, p. 16.

6. WOLMAN, op. cit., pp. 168-169.

7. JACKSON, R. I. "International Fisheries and Marine Pollution," International Conference on Oil Pollution of the Sea, Proceedings (Conference held in Rome, October 7-9, 1968), pp. 30-31.

8. BUSCH, D. D. and MEARS, E. I. "Ocean Pollution: An Examination of the Problem and an Appeal for International Cooperation," San Diego Law Review, No. 7 (July 1970), p. 601.

9. SCHACHTER, and SERWER, op. cit., pp. 110-111.

10. "Legal Control of Water Pollution," University of California,
    Davis, Law Review, No. 1 (1969), pp. 173-174.

11. LESTER, A. P. "River Pollution in International Law," American
    Journal of International Law, No. 57 (October 1963),
    pp. 828-853.

12. O'CONNELL, D. M. "Continental Shelf Oil Disasters: Chal-
    lenge to International Pollution Control," Cornell Law
    Review, No. 55 (November 1969), p. 122.

13. LEGAULT, op. cit., pp. 1-6.

14. SWEENEY, J. C. "Oil Pollution of the Oceans," Fordham Law
    Review, No. 37 (December 1968), p. 208.

15. MENDELSOHN, A. I. "Maritime Liability for Oil Pollution -
    Domestic and International Law," George Washington
    Law Review, No. 38 (October 1969), p. 28.

16. SWEENEY, op. cit., pp. 187-188.

17. CLINGAN, T. A., Jr. and SPRINGER, R. "International Regu-
    lation of Oil Pollution," working paper for the Interna-
    tional Law Panel of the President's Commission on Ma-
    rine Science, Engineering, and Resources (undated),
    pp. 28-29.

18. United States Senate Committee on Commerce. Environmental
    Activities of International Organizations. Washington:
    U.S. Government Printing Office (1971), p. 22.

19. CLINGAN, and SPRINGER, op. cit., pp. 26-32.

20. SHUTLER, N. D. "Pollution of the Sea by Oil," Houston Law
    Review, No. 7 (March 1970), p. 430.

21. The Convention on the Continental Shelf, done at Geneva,
    April 29, 1958 and in force in 1964. See 499 United
    Nations Treaty Series, p. 311.

22. O'CONNELL, op. cit., p. 124.

23.   The Convention on the High Seas, done at Geneva, April 29,
       1958 and in force in 1962. See 450 United Nations
       Treaty Series, p. 82.

24.   YOSHIOKA, T. T. "Problems of International Control of Oil
       Pollution of the Sea," paper presented to Professor
       Rodgers' law seminar, University of Washington (un-
       dated), p. 14.

25.   The Convention on the Territorial Sea and the Contiguous Zone,
       done at Geneva, April 29, 1958 and in force in 1964.
       See 516 United Nations Treaty Series, p. 205.

26.   See International Legal Materials, vol 9 (1970), p. 25 for the
       complete text of the Convention.

27.   O'CONNELL, D. M. "Reflections on Brussels: IMCO and the
       1969 Pollution Conventions," Cornell International
       Law Journal, No. 3 (Spring 1970), p. 177.

28.   LEGAULT, op. cit., pp. 7-8.

29.   O'CONNELL, 1970, op. cit., p. 167.

30.   See International Legal Materials, Vol. 9 (1970), p. 45 for the
       complete text of the Convention.

31.   Ibid., Civil Liability Convention, Article 5, Paragraph 1.

32.   HEALY, N. J. "The International Convention on Civil Liability
       for Oil Pollution Damage, 1969," Journal of Maritime
       Law and Commerce, No. 1 (January 1970), p. 322.

33.   A. I. Mendelsohn and Robert H. Neuman discuss the merits and
       deficiencies of the Civil Liability Convention in
       hearings before the United States Senate Air and Water
       Pollution Sub-Committee of the Senate Committee on
       Public Works, July 21 and 22, 1970.

34.   HEALY, N. J. and PAULSEN, G. W. "Marine Oil Pollution
       and the Water Quality Improvement Act of 1970,"
       Journal of Maritime Law and Commerce, No. 1 (July
       1970), pp. 537-572.

35. International Convention on the Establishment of an International Fund for Compensation for Oil Pollution Damage, December 18, 1971, manuscript copy, 42 pp.

36. Canada, Parliament, House of Commons. Order 209 (November 12, 1971), and Aide-Memoire to United States Government (August 18, 1971), Appendix II.

37. See International Legal Materials, Vol. 9 (1970), p. 1 for the complete text of the Convention.

38. See International Legal Materials, Vol. 8 (1969), p. 497 for a complete text of the Contract.

39. O'CONNELL, 1970, op. cit., pp. 183-184.

40. MENDELSOHN, op. cit., pp. 7-9.

41. See International Legal Materials, Vol. 10 (1971), p. 137 for the full text of the Contract.

42. HARRIS, M. B. and LOVETT, A. "Recent Developments in the Law of the Sea: A Synopsis," San Diego Law Review, No. 7 (July 1970), p. 656.

43. HEENEY, A. D. P. and MERCHANT, L. T. Canada and the United States: Principles for Partnership. Ottawa: Queen's Printer (1965), p. 1.

44. "IJC Asked to Study Pollution Risks From Lake Erie Oil Spills," Department of State Bulletin, No. 60 (April 7, 1969), pp. 296-297; International Joint Commission. Special Report on Potential Oil Pollution, Eutrophication and Pollution from Watercraft. Ottawa and Washington, D.C.: International Joint Commission (1970)

45. ABLETT, D. "U.S.-Canada Near Pact on Cleaning Up the Great Lakes," Vancouver Sun, June 9, 1971, p. 16.

46. Ibid.

47. United States, Executive Office of the President, Office of Science and Technology. The Oil Spill Problem. Washington, D.C.: U.S. Government Printing Office

183

(1968), p. 17; United States, Departments of Interior
and Transportation. Oil Pollution – A Report to the
President. Washington, D.C.: U.S Government Printing Office (1968), p. 27.

48.   BLOOMFIELD, L. M. and FITZGEARLD, G. F. Boundary Water
        Problems of Canada and the United States. Toronto:
        Carswell (1958), p. 15.

49.   CLINGAN, and SPRINGER, op. cit., pp. 30-31.

50.   BLOOMFIELD, and FITZGEARLD, op. cit., p. 17.

51.   International Joint Commission. Rules of Procedure and Text of
        Treaty. Washington, D.C.: U.S. Government Printing
        Office (1965), p. 17.

52.   BLOOMFIELD, and FITZGEARLD, op. cit., p. 17.

53.   Personal communication, J. L. MacCallum (May 8, 1970).

54.   JORDAN, F. J. E. "Recent Developments in International Envi-
        ronmental Pollution Control," McGill Law Journal,
        No. 15 (1969), pp. 283-285.

55.   Personal communication, J. L. MacCallum (May 8, 1970).

56.   REMPE, G. A., III. "International Air Pollution – United States
        and Canada – A Joint Approach," Arizona Law Review,
        No. 10 (Summer 1968), p. 144.

184

CHAPTER 5
AN ALTERNATIVE ORGANIZATION FOR OIL POLLUTION
PREVENTION AND CONTROL

International treaties and domestic laws in and between Canada
and the United States do not provide adequate protection or compensation
for victims of transboundary oil pollution. Chapters Three and Four documented the strength and weaknesses of the appropriate legislation at various areal scales including state, provincial, national, bilateral and multilateral laws. The problem of transboundary oil pollution necessitates an
international response in addition to the improvements which can be fashioned through domestic legislation. Various alternatives are available,
including improved domestic legislation discussed earlier; however, regional arrangements, while somewhat utopian in concept, offer the most
realistic, comprehensive solution to the problem in the study area.

PRINCIPLES NEEDED TO PREVENT AND CONTROL
OIL POLLUTION

Existing institutions have addressed themselves to some of the basic issues involved in removing the externalities created by spilled oil.
Ideally, any new organization would incorporate the following principles
if its aims were to require polluters creating technological externalities to
internalize all costs associated with the spill. The public would have a
recognized right to a pollution free environment, a right which would en-
able persons to take legal actions on the public's behalf for any damage to
the public environment. Oil transport and drilling would be recognized
explicitly as ultra-hazardous activities and regulations would be passed
expressly forbidding the discharge of oil. Owners of oil and operators of
vessels would be jointly and severally liable for all oil damage, and the
burden of proof would be on the polluter. Responsibility for cleanup
would rest primarily with the polluter under supervision from the proposed

organization; however, the organization would have to have at its disposal the full range of equipment and materials needed to respond to a large
undetected international spill. Claims against the polluter would be valid for direct and indirect damage.

Any organizations seeking to prevent and control oil pollution
would have to have adequate power and resources to respond to the largest of spills. Such an organization would have to be capable of delineat-
ing those responsible for oil spillage, responding to the crisis, and developing the means of enforcing the principles described above. The provision of adequate fiscal resources will be essential if any new organization
is to institute a program of prevention and control, as well as compensate
innocent third parties. Revenue for such an organization could be provided by a regional fund generated from a tax on the number of tons of oil
brought into or transported within the region.

Many lawyers have contended that ship and cargo owners should
face absolute liability for spillage. They argue that the transport of oil is
a hazardous activity, that oil production and transportation is financed
through global corporations, and that industry is therefore better able to
absorb and distribute the costs. Others argue for strict liability with no
limits, but grant that exceptions should rightfully be made for acts of God
and negligence on the part of governments. They maintain that no corporation should be penalized because of actions which are clearly beyond its
control. Most industry representatives have resisted attempts to establish
absolute or strict liability and have argued for limited liability with an
upper ceiling on the total liability in any one incident. If all externalities are to be removed from oil spillage, there can be no limit to the amount
of funds available for cleanup and compensation. Absolute liability would
provide sufficient funds, but it might serve to drive small operators out of
business because they could not secure insurance for unlimited amounts.
Strict liability would generate sufficient revenue; however, the exceptions

186

noted above could create situations in which third party damage claims would be difficult to sustain. Limited liability has an obvious limitation regarding damages in excess of the total amount of liability.

One solution to the problem might be to create a special regional fund in addition to the laws of strict and limited liability which exist in the various jurisdictions in the study area. An international fund could be created, based on the amount of oil carried in ships and barges within the region. All operators carrying oil would pay a specified rate per ton. The rates would be based on the type of equipment and safety record of the company. Thus those operators with poor spillage records and faulty equipment would be forced to pay a higher tax for the privilege of carrying oil in Puget Sound and the Strait of Georgia than operators with good safety records and modern equipment. This fund would be used in addition to the liability of ship and cargo owners, would be independent of strictly national oil spills, and would serve as a more equitable and efficient means of pollution control and compensation. Under such a system the damaged party, upon presentation of adequate proof of damage, would be compensated fully, without having to resort to suits against the polluter. Many of the complicated questions of negligence, trespass, and liability would not have to be litigated. Innocent parties would be assured of immediate compensation for damages they have suffered. Consumers would be forced to pay increased costs for their petroleum products, but such costs would be more equitably distributed amongst the oil users than at present. Moreover, general revenue from governments would not have to be used to finance prevention and compensation operations for a particular industry. By using a fund in addition to the liability, small operators, especially some of those transporting oil within the region, would be able to secure insurance and remain in the industry, and, at the same time, the public would be fully protected. These small operators, along with the owners of large tankers, will be encouraged to improve ship construction and design and therefore benefit from a lower tax per ton on the amount of oil transported. One of the most important functions of any international commission which might be established to manage the oil pollution problem in the region will be to strike a tax rate which generates sufficient revenue and gives owners an incentive to improve the design and construction of their vessels to minimize pollution. The fund created by such a tax could also be used to finance the operations of such an international commission.

THE AREAL SCALE OF CONTROL

While many authors favor a regional approach to pollution control, few agree on the areal scale at which regional controls should be applied. Hovanesian and O'Connell place primary emphasis on developing an international framework within which coastal states might take coordinated actions to control pollution. O'Connell envisions IMCO taking a broader role as an objective fact-finding body, but argues it would be utopian to suggest that such an agency could secure enforcement power for its decisions, given the divergent interests of flag and coastal states. Creation of an international fund for pollution control would serve as useful security for private insurance coverage, but O'Connell recognizes that such a fund would require "near universal adherence to the system to prevent exploitation by non-members."[2] Hovanesian similarly recognizes that "effective relief would have to be world wide in scope and orientation if it were successfully to transcend traditional barriers of conflicting legal systems and nation legislation, jurisdiction and other intricate man-made obstacles."[3] Both authors are cognizant of the necessity of transcending national boundaries, but they fail to assess the difficulties of administering a fund or program if universality were not achieved. States who were not members of such a fund would not be required to ensure that compensation was paid for oil spilled by their nationals. In the absence of firm rules between all nations there can be no guarantee that all externalities will be removed.

Others argue that regional arrangements should be concluded on a smaller scale taking into account existing international agreements but such arrangements should, if necessary, be independent of existing international standards. Nanda, Busch and Mears[4] believe that regional arrangements between industrialized nations will facilitate more rapid action against pollution. Such arrangements would take the form of international bodies whose functions would be to standardize and determine: what pollutants would be subject to control; the scope and extent of liability; and preventive and remedial efforts to be taken. These bodies would be essentially investigating agencies, and would operate very effectively (the authors believe) in the Baltic, Mediterranean, North Sea, and along the Canada-United States border.

Jordan and Rempe have given particular attention to pollution problems along the Canada-United States border.[5] Jordan argues that the establishment of a supranational pollution control authority is essentially utopian and that neither Canada nor the United States would vest broad powers in an international agency, whether it be the International Joint Commission or some other body.[6] Even taking into account this political fact, real pollution problems do exist along the border which will, nevertheless, require international cooperation.

It is not the purpose of this study to ignore what needs to be done, merely because the political realities may preclude immediate implementation of the recommendations. The recommendations which follow are a comprehensive beginning to solving the problem of oil pollution in Puget Sound and the Strait of Georgia, and are based on the belief that a regional bilateral commission between Canada and the United States is the best agency and the most appropriate scale at which to tackle the problem. Implementation of the entire program is one way to remove most of the externalities that would be created by an international oil spill; implementation of various parts will be less than optimal but probably an improvement over existing international arrangements in the region.

AN ALTERNATIVE ORGANIZATION

The problem of oil pollution in Puget Sound and the Strait of
Georgia can best be solved through establishment of a joint international
commission between Canada and the United States, which should be based
on the principles discussed earlier in this chapter. The joint international
commission approach is not new in international pollution control, but, as
Lester has noted, it has been used primarily only to deal with international river basin development and use.[7] Such an approach has now gained
considerable acceptance, and the International Law Association recognizes the importance of joint commissions in seeking solutions to international river basin problems.[8] Applying the joint commission concept to
salt waters, especially those which have many of the characteristics of
fresh water estuaries, is a logical extension, even if such an extension
still leaves unsolved the pollution problems of the high seas which are
subject to multilateral rather than bilateral control. Moreover, Canada
and the United States already vest considerable power in a joint commission in Puget Sound and the Strait of Georgia.[9] The Pacific Salmon Commission, first organized to protect sockeye and later pink salmon of the
Fraser River system, regulates fishing times for Canadian and American
fishermen. Fish, like spilled oil, are a common property resource, and
there is a need for international regulations when ownership cannot be
internalized within one political jurisdiction. The fish of the Fraser River
system migrate through American waters in Puget Sound as they wind their
way to the spawning areas which are all in Canada. Since each nation
has the capability to annihilate the runs, some agreement was needed.
No scientific data or economic rationale were advanced in favor of the
present 50-50 catch division, but an unequal apportionment would have
created an unwanted source of friction between the two countries. Similarly, in the Columbia River Treaty, a 50-50 division was accepted in

allocating downstream power benefits. Thus there is precedent in international attempts at pollution control and in relations between Canada and the United States in the study area for the establishment of a joint international agency to prevent and control oil pollution.

Establishment of such an agency could be accomplished in either of two major ways. It could emanate from a separate treaty concluded between Canada and the United States, or it might be possible to add a clause to the Boundary Waters Treaty which already deals with fresh water problems. In either case, establishment of such an agency, based on the principles noted above, will require fundamental changes in the approach to pollution control in both countries. Not only would national governments have to surrender powers to the commission, but the same governments would require the acquiescence and support of various levels of government at the subnational level.

The commission need not be radically different in structure from existing agencies between the two countries. There could be an equal number of commissioners appointed by member governments as in the six member International Joint Commission. Commissioners would be responsible for overall policy, but the day to day affairs of the commission could be run by an executive director, as is the case with the Pacific Salmon Commission. There would be a research branch associated with the commission and the commission would either have its own enforcement officers or rely on enforcement officers from the civil service of the national, state, and provincial governments. The major radical departure would be in the method of financing the commission, and in the powers and functions assigned to the commission.

One of the major functions of the commission would be the development and institution of specific measures to prevent oil pollution. This might take several forms. Specific requirements on ship construction (including compartmentalization and double hulls) could be required of all ships entering the study area, or those without such features could be forced to pay a higher tax on the amount of oil carried. As commercial traffic increases and the size of tankers continues to grow, it might become necessary to provide well defined routes and positive traffic control from a central body such as the proposed commission. Nanda comments:[10]

> ... the lack of any obligation on the master of a ship to report his course, speed and position to coastal stations should be replaced by control systems similar to the aircraft control systems. Present technology, using LORAN type radar equipment with identification attachments, long range, shore based radar equipment or Doppler systems could perhaps efficiently track all vessels from port to about 500 miles offshore. Computerized control could be established, and ships approaching ports could be regulated as aircraft are when they approach metropolitan airports. This device could be used to warn ships of collisions or navigational errors that might cause disaster on congested sea routes.

Traffic control by the proposed commission might be further developed by regulating and segregating dangerous cargo ships such as oil tankers from the main traffic. Consideration might also be given to excluding commercial ships from certain areas of Puget Sound and the Strait of Georgia. Areas where the ecology is very sensitive might be permanently protected from commercial transport. Other areas might be restricted during sensitive seasons. When the salmon runs are proceeding through the region, ships carrying dangerous cargoes might have to be excluded. The commission might also investigate the dredging and removing of navigational hazards on the sea lanes where the damage to the environment resulting from grounding or collision would be greater than that done in improving the sea lanes. The position of terminals at the ends or along the sea lanes would come under the purview of the commission. Future sites for petroleum refining and transfer should be restricted to areas having low ecological value and natural containment, such as an already polluted bay with little mixing current. The commission might require that such sites be as close as possible to sea to minimize the time spent in the region and to reduce the risk of collision or grounding while entering or leaving. Thus, the commission might require Alaska oil destined for Puget Sound, for example, to be discharged near Port Angeles and then transferred by pipeline to other areas within the Sound. If ships do come into the region, the commission might set standards of minimal performance and, if these are not met, require the ships to be escorted by tugs.

Integrated international cleanup plans should be developed by the commission and, for spills occurring in domestic waters, should be closely associated with national efforts. The commission may want to develop its own response capability, or it may rely on the facilities available from national governments. In either case, the commission should follow the call set forth by the International Joint Commission in their study of oil pollution in the Great Lakes. The International Joint Commission urged the two governments to arrange for the development of a "coordinated international contingency plan so that both countries may quickly and effectively respond to major accidental spills of oil or other hazardous materials in the boundary waters of the Great Lakes system."[11]

The commission should require ships to have means of preliminary control on board. Regulations would be promulgated establishing the number of men necessary for an emergency crew on board ship, even if the ship were automated and did not require such a crew for normal operations. In such an instance a team of specialists might be made available to board the ship with the pilot before the ship entered the region. Ships would also be required to have independent, deck-mounted auxiliary pumps and heaters for tankers, along with a universal fitting to which others might attach to withdraw oil from sunken vessels. Acceptable booms would also be necessary on board ships so personnel could respond quickly and take initial preparations to control the spill. The commission would have available at stations along major sea lanes large amounts of heavy weather booms, absorbants, skimmers, pumps, and heaters. All such equipment would be aboard barges or other facilities, and thus would be highly mobile. The commission would be responsible for hiring such manpower as is necessary to clean up major spills. Monies necessary to pay for such equipment and manpower would come from the regional oil pollution fund.

Liability and compensation resulting from oil spills would be based on principles discussed above. In addition to these after-the-fact functions, the commission would be the body designated to supervise the type and cost of restoration. The commission would have power to suspend masters and pilots involved in any pollution incident, and to conduct an investigation to determine further action against those involved in the incident.

Creation of such a commission would go a long way towards internalizing the externalities associated with spilled oil along the international boundary. Establishing a flexible tax with lower rates for acceptable preventive measures would cause potential polluters to assess more realistically the transport and damage costs. No commission or tax will eliminate oil pollution, but they will help to minimize potential damage to the ecological integrity of the region. Moreover such a tax may serve to achieve the same results for non-recurring pollution that a system of charges based on a non-linear damage function serves for recurring pollutants from industrial processes.[12] The difficulty in instituting such a tax will be to incorporate ecological consequences, irreversibility, and non-linear damages, in rates that are based on the probability of spillage from various oil pollution sources.[13] If this can be accomplished, and the commission is able to reduce oil pollution and provide adequate compensation, then the commission might serve as a useful model for controlling oil pollution in other areas.

APPLICATION OF THE COMMISSION APPROACH TO
OTHER AREAS AND RESOURCES

The International Joint Commission has already investigated the

problem of oil pollution in the Great Lakes. Other areas along the inter-

national boundary also face potential oil pollution problems. The most

prominent of these is the potential problem along the Maine-New Bruns-

wick border. Crude oil (destined for Portland, Maine, and later Mont-

real) comes close to this border, and the tides of the Bay of Fundy are

capable of moving spilled oil onto the beaches of New Brunswick and

Nova Scotia. In addition, proposals have been made to develop oil re-

fineries in Maine because of the presence on the northeast coast of large

deepwater ports which can handle some of the largest supertankers. Many

of the proposals have been stalled or shelved, but the danger of spilled

oil from any of these facilities could pose a very serious international

problem.[14] The same type of commission which has been proposed for the

Puget Sound and Strait of Georgia area could prove useful in meeting oil

pollution problems in both the Great Lakes and the Portland, Maine area,

although some of the unique physical problems and the greater number of

political jurisdictions involved could call for modifications in the struc-

ture and organization of the commission.

Oil pollution is only one of several resources and pollution inci-

dents causing localized problems along the international boundary. Many

of the water problems have been investigated by the International Joint

Commission and, more recently, this organization has given attention to

air pollution. Attempts have been made to suggest solutions for particu-

lar problems, such as air pollution in the Windsor-Detroit area. The in-

ternational commission could well be used as a model for the control of

air pollution. The physical features of air pollution are far different, but

the principle of internalizing externalities within one jurisdiction is,

conceptually, a problem common to both international air and oil pollution. A comprehensive approach to pollution problems, incorporating principles suggested earlier, is fundamental to the solution of international boundary problems between Canada and the United States. Indeed the proposal for an international commission to solve one particular problem could be very shortsighted. It may be that commissions might be established in several problem areas along the boundary to deal with all resources and transboundary problems in that region. Thus, in the Puget Sound and Strait of Georgia region, such a commission might be able to deal effectively with such thorny issues as air and oil pollution, migrating salmon, and the Skagit Valley-Ross Dam issue. Within this framework, the commission would not replace the International Joint Commission; it would be the local body dealing with special regional problems. The International Joint Commission would supervise local bodies and establish policies where it could be clearly shown that the resource problems were common to several areas along the Canada-United States boundary.

The commission approach could also have applications outside of Canada and the United States. Earlier sections of this study have referred (rather obliquely) to the problems of oil pollution along the coasts of several nations, and, more particularly, in the Mediterranean, Baltic, North Sea, and English Channel areas. This pollution stems partly from oil drilling, but is more closely linked to the maritime trade in oil between Europe and the oil producing areas of the Middle East. Agreements concerning oil in these areas have served to facilitate exploration and exploitation, but little has been accomplished on a regional scale to prevent and control pollution. Nevertheless, contacts have been established and conferences have been held to plan responses to the threat of oil pollution. Because of the number of nations involved and the complexity of shipping interests, rapid agreement on oil pollution prevention is not likely to emerge. Positive control of all shipping has been proposed for the English Channel and Persian Gulf areas and this could lead to broader

196

agreements on contingency plans and compensation. Because of the vast areas involved and the proximity of some of the regions mentioned above, the comprehensive approach proposed for Puget Sound and Strait of Georgia area may be utopian and premature for these areas. However, implementation of features, singly or in combination, such as regional control of shipping, common standards for ship construction, compensation for direct and indirect damage, would immeasurably help reduce the potential for damage and the impact of externalities from the oil industry.

This chapter has argued that the problem of oil pollution along the international boundary between Canada and the United States cannot be met solely by domestic legislation or present international arrangements, and that a new and different type of international response is urgently needed. The simplest international response is the bilateral agreement. Undoubtedly any bilateral agreement which might emerge on this problem would reflect the limitations of domestic legislation on oil pollution and the economic and political problems along the international border. There is, however, a need to outline an optimal solution to oil pollution problems in the region, based on a more equitable distribution of the costs involved. An earlier part of this chapter argued that this can be achieved most simply by an international commission reflecting regional needs. It is also submitted that a successful program would not retard progress to more effective global controls, but would serve as an example of what might be done in other areas, while meeting the current threat to Puget Sound and the Strait of Georgia more effectively. This is not to suggest that such a program would be implemented immediately and in its entirety by the two governments. Profound changes in the attitude of both Canada and the United States towards pollution would be necessary. Such fundamental changes in attitude would involve recognizing: the international nature of the problem; the need to foster an ecological conscience; the need to develop institutions that have power to implement this viewpoint;

197

the need to prevent crises before they arise; and the need to surrender domestic power to an international organization. Nevertheless, implementation of sections of the comprehensive approach outlined above would go a long way to meeting the current threat of international oil pollution damage in Puget Sound and the Strait of Georgia.

REFERENCES

1. HOVANESIAN, A., Jr. "Post Torrey Canyon: Toward a New
    Solution to the Problem of Traumatic Oil Spillage,"
    Connecticut Law Review, No. 2 (Spring 1970),
    pp. 632-647; O'CONNELL, D. M. "Reflections on
    Brussels: IMCO and the 1969 Pollution Conventions,"
    Cornell International Law Journal, No. 3 (Spring 1970),
    pp. 161-188.

2. Ibid., p. 187.

3. HOVANESIAN, op. cit., pp. 644-645.

4. NANDA, V. P. "The Torrey Canyon Disaster: Some Legal
    Aspects," Denver Law Journal, No. 44 (Summer 1967),
    pp. 400-425; BUSCH, D. D. and MEARS, E. I.
    "Ocean Pollution: An Examination of the Problem and
    An Appeal for International Cooperation," San Diego
    Law Review, No. 7 (July 1970), pp. 574-604.

5. JORDAN, F. J. E. "Recent Developments in International Envi-
    ronmental Pollution Control," McGill Law Journal,
    No. 15 (1969), pp. 279-301. REMPE, G. A., III.
    "International Air Pollution – United States and Canada –
    A Joint Approach," Arizona Law Review, No. 10
    (Summer 1968), pp. 138-147.

6. JORDAN, op. cit., p. 300.

7. LESTER, A. P. "River Pollution in International Law," American
    Journal of International Law, No. 57 (October 1963),
    pp. 828-853.

8. JORDAN, op. cit., p. 298.

9. I am indebted to Professor Ralph W. Johnson of the School of Law,
    University of Washington, for noting this precedent.

10. NANDA, op. cit., p. 422.

11. International Joint Commission. Special Report on Potential Oil

Pollution, Eutrophication and Pollution from Watercraft. Ottawa and Washington: The International
Joint Commission (1970), p. 28.

12.   See Chapter Two which calls for a re-ordering of priorities in
       allocating scarce resources.

13.   BROWN, G., Jr. and MAR, B. "Dynamic Economic Efficiency
       of Water Quality Standards or Charges," Water Re-
       sources Research, No. 4 (December 1968), pp. 1153-
       1159; PARKER, D. S. and CRUTCHFIELD, J. A.
       "Water Quality Management and the Time Profile of
       Benefits and Costs," Water Resources Research, No. 4
       (April 1968), pp. 233-246.

14.   McDONALD, J. "Oil and the Environment: The View from
       Maine," Fortune, No. 83 (April 1971), pp. 84-89,
       146-147, 150.

CHAPTER 6
CONCLUSION

> The simple truth is that we are far from one world
> politically. But, by necessity if not by choice,
> we are one world environmentally. States have
> sovereign rights — but so do people. We cannot
> rely on the political habits of the past to save our
> environment for the future.[1]

This study is indirectly concerned with the management of all global environmental problems. The central issue raised by the study is whether existing national and international institutions are capable of internalizing all relevant costs associated with major international oil spills affecting coastal areas. Marine pollution is, of course, but a part of the totality of global environmental problems which confront us today. The study also offers perspectives on the nature of the governmental systems operating in the study area and on the types of measures needed to deal more realistically with pollutants such as spilled oil.

The inescapable conclusion is that existing institutions are either so powerless that they are unable to reduce the impact of international oil pollution or they lack the specific jurisdiction to do so. Approaches to oil pollution control have traditionally focused on prevention, control, or compensation, but seldom has a single organization performed all functions. Chapter Five argued that a regional commission with broad powers, free of traditional national constraints, would go a long way toward meeting the crisis at hand, and that even incremental improvements would reduce the threat from spilled oil, provided that these improvements were established on a regional basis. Chapter Three noted the desirability of improvements in domestic law at all levels, and the need for strict penalties in the absence of a bilateral agreement between Canada and the United States over oil pollution control in the region. The particular recommendations and the lessons learned from this study are relevant to the prevention and control of all pollution at the international level. While oil pollution has been the subject of this study, other pollutants contribute to a complex series of events which cause deterioration in the quality of the world environment. The solution to this complex problem lies in a comprehensive approach which examines the multitude of complexities and subtleties involved. A merely piecemeal study focusing on one particular problem will be inadequate. Nevertheless, the difficulties of achieving agreement among nations who do not view the problem of oil pollution as a global crisis dictate a more modest approach, such as the commission approach suggested above. Schachter and Serwer place the problem in perspective:[2]

> . . . we need a many-sided institutional approach to achieve the right balance. Pollution problems will not be solved by a single discipline, a single institution or a single wave of enthusiasm. Science can provide certain types of information, but that information will have to be communicated effectively to the international and national decision makers. There is certainly a need for new institutions, though a large part of the solution will lie also in making old institutions more effective.

The nature of the problem in the study area is revealing because the particular crisis has not been communicated effectively to the national governments and has only been regarded as an international problem by limited segments of the population. The public and government of British Columbia and preservationist groups in Washington State certainly regard the potential spillage problems as an international issue requiring a response from both Canada and the United States. These groups, influenced to a large extent by the general attention given oil pollution by the media, have defined the potential problem as a crisis because they have seen the consequences of major spills in other areas and witnessed the effect of smaller spills at Bellingham and Anacortes.[3] Other groups, especially those concerned with the depressed condition of the Washington State economy, have argued that increases in the amount of oil entering the region will help reduce unemployment and will contribute to the financial recovery of the state. To these individuals the potential deterioration of environmental quality in the region from oil pollution is not a crisis. They are more concerned with devising programs which will help the economy and are willing to accept the environmental hazard in pursuit of what they consider more important economic goals. With reference to this attitude, it is interesting to speculate whether the people and government of British Columbia would be so prepared to recognize the environmental crisis if the same economic conditions existed in British Columbia and if they could see economic benefits accruing to the province. It is apparent, at any rate, that not the entire populace of the study region even believes a crisis exists. The unanimity of opinion in British Columbia has forced the Canadian government to initiate discussions with the United States, and the Secretary of State for External Affairs has consulted the United States Secretary of State. The United States government, however, has been largely unresponsive. It does not see a particular threat of oil pollution from Canada in the region, and the majority of pressure placed on Washington, D.C., has favored expanding the oil industry in the state. Even if there were unanimity of opinion within the region that an environmental crisis existed, pressure from the oil industry in the United States might convince Washington not to intervene.

The role of crisis as a tool for accomplishing radical change has obviously been muted in the study area. Svart has commented that "some level of impaired environmental quality is a necessary condition of a perceived crisis."[4] But he could also have added that the perceived crisis can be generated by the consequences of similar problems in other areas. Certainly the difficulties associated with the Torrey Canyon and the Arrow have had a substantial influence on public opinion in British Columbia and Washington. Perceived crises, however, seldom move governments to action, especially if the crisis is regionally centered. The environmental crisis over potential oil spillage in Puget Sound and the Strait of Georgia may be very real, but, as noted earlier, solutions to the spillage problem require a national response. National governments have been unwilling to take action when there is not unanimity about a perceived crisis. A profound gap can exist between those who may see a crisis in a small localized area and those at the national level who have the power to initiate corrective measures to solve international pollution problems. This gap is not likely to be bridged until there is unanimity of opinion about the perceived crisis, and until the region is able to make its political weight felt. Moreover, the problem in the study area will not be solved until both Canada and the United States recognize the potential catastrophe. Opinion about the perceived crisis in Washington State does not generate optimism, but should an actual catastrophe occur in the area (and especially if this catastrophe were centered in Canadian waters), regional preventive measures might be instituted by the national governments.

The spillage problem has been compounded by a failure to consider a broad range of alternatives at the regional, national, and international levels. Demand for petroleum in the study region is expected to parallel that in other areas, and projections indicate that one or two additional refineries may be needed in both Washington and British Columbia before 1980. Oil companies, especially those in Washington State, have assumed that this demand will be met with oil from Alaska. Little or no consideration has been given to using other sources of oil. The Cherry Point Refinery, for example, will initially use Alberta crude; yet this supply will not be used as a permanent source. If other alternatives have in fact been considered by the oil companies or by the governments, such discussions have not been made public, and no hearings have been called to discuss the subject. In British Columbia, refined products continue to be transported to Vancouver Island by tanker and barge with no public consideration being given to the feasibility and practicality of a products pipeline. This failure to publicly consider alternatives at the regional level is symptomatic of existing apathy. The problem is magnified considerably when discussions are held on the exploitation and transportation of Alaskan oil. Questions being asked by the Department of the Interior, for example, consider some of the national implications, but (as this study has demonstrated) the international environmental impact has been largely ignored. There is a need to ask questions and solve problems at the national and international levels if a rational assessment is to be made. In the Cherry Point case described above, such an assessment might begin by asking the following questions: (1) Do we need oil? (2) Do we need Alaskan oil? (3) Could alternative supplies meet the demand? (4) Do we need Alaskan oil now? (5) If we need Alaskan oil, what are the alternatives for transporting it to markets? (6) Are some areas so ecologically sensitive that further development of the oil industry should be restricted? (7) If we restrict development in certain areas such as Alaska, how do we compensate the region? (8) If we permit development, how do we protect the environment which would be threatened by development?

Such questions obviously involve Canada as well as many other nations, yet the decision on exploitation and method of transportation will likely be made at the national level. Confronted by this reliance on national decision making and sovereignty, application of international management of resources is limited. There is little chance that Firey's priorities in allocating resources can be implemented. This is not to argue that involving more nations in what has heretofore been regarded as

PLATE 7a
Environmental disruption caused
by pipeline laying

PLATE 7b
Mackenzie Valley Pipe Line
Research Limited's experimental
facilities near Inuvik

a domestic decision will automatically ensure that the proper questions will be asked or that solutions will be found. International decision making is fraught with difficulties. Canadian concerns about a continental energy package, United States fears of relying on foreign oil sources, and Middle East demands for higher crude oil prices would complicate such decisions. Failure to recognize the international environmental impact of national resource use decisions will result in a grave misallocation of resources. More consideration should be given to opening the consideration of alternatives to closer public scrutiny.[5] There is a demonstrable need for an organized international political constituency with an ecological conscience, capable of influencing national decisions.

The failure of national governments to consider broad alternatives raises the question of whether national solutions can protect the quality of the international environment or whether such protection can only be secured through multilateral agreements on a global scale. The study has shown it is possible for national and even subnational governments to take stringent action beyond that existing at the international level. The Washington Oil Spill Act of 1970 and the Canada Shipping Act of 1971 demonstrate this point well. The study has also shown that existing multilateral agreements give too great an emphasis to the rights of flag states and to freedom of the seas. It is evident that new approaches are needed to prevent and control oil pollution, approaches that take into consideration the broader international impact of any decision.[6] National legislation has often provided the basis for international action. The introduction of TOVALOP and subsequent modifications of that agreement were attempts to stave off higher liability limits for oil pollution in Britain and the United States. Amendments to the Canada Shipping Act (noted above) and passage of the Arctic Waters Pollution Prevention Act may be a prelude to broader international controls. This study has noted the desirability of broad international action and global agreement, but it has also demonstrated that unilateral national actions need not be contrary to the international environmental interest. As long as the international community retains its alliance with oil and transportation interests, national legislation will be needed to prod that community to more stringent controls.

The international community should give first priority to the question of ownership of the seas. Chapter Two noted the difficulties involved in managing a common property resource, especially where ownership of the resource is unclear or absent. Global marine law is not well established and innovative solutions therefore are perhaps more possible in this area than with problems such as land and fresh water pollution where laws and control institutions are comparatively well developed. Little can be done to solve oil pollution of the sea, however, until property rights are established. Ideally ownership rights would be entrusted to some international groups, a solution which has been suggested for allocating the resources of the sea bed. Such a solution would avoid a new round of global conflict over dividing up the oceans into national areas and thus prevent even greater problems of transboundary pollution.

Rapid changes in existing laws to prevent and control oil spillage are further complicated by the dissimilar natures of the federal systems of Canada and the United States. The United States faces few jurisdiction problems in controlling oil pollution. Both the state and federal governments have authority over oil spillage in state waters, and have coordinated their planning and contingency operations. In the Canadian case, however, there is a divided jurisdiction. The federal government has control over navigable waters, but damage sustained above the low water level or in other areas of provincial responsibility would be beyond its jurisdiction. A joint federal-provincial agreement is needed to solve the impasse. The provincial government has seen fit to let Ottawa take care of oil pollution in federal waters, and refuses to recognize responsibility for its own shores and waters. Whatever the political motives for such action, whether these be a desire to avoid duplication of effort, a desire to avoid large expenditures on oil pollution prevention, or a move to force the federal government to reconsider its position on offshore drilling, the consequences are clear. Because of this divided jurisdiction, Canada has found it more difficult than the United States to implement comprehensive changes in oil pollution controls that apply to all coastal waters and shores.

Should a joint federal-provincial agreement be consummated, the Canadian government would likely be able to implement programs more quickly than the United States, if the pollution problem were recognized as a crisis worthy of government priority. The Canadian federal government is responsible to the legislative branch, and it is able to secure quick passage of legislation by reason of the majority it commands in parliament. Committees of the House of Commons have limited powers to amend legislation, and government regulations have not been subject to broad public examination before promulgation. Programs of the United States President, on the other hand, can be delayed, stalled, or defeated by one of the congressional committees. The committee system, with its emphasis on seniority and its fragmented approach to environmental issues, only serves to delay and frustrate the legislative process. In addition, Congress has inadvertently surrendered its leadership on environmental issues to the executive branch. The essential contrast between decision making in each federal system is that the Canadian government is capable of dealing with environmental problems quickly (given agreement on the jurisdictional problem), while the United States government is hampered by the more involved legislative process. This comparison does not necessarily place a value judgement on the merits of either system. Together with the lack of action in response to a localized but important international oil pollution problem, the failure of each national system to deal adequately with the oil pollution threat leads irrevocably to two important conclusions. First, reforms are needed in each system (be it on the jurisdictional issues or the legislative process) to meet the growing magnitude and scope of environmental problems. Second, more extensive research is needed into the differing nature of decision making in each federal system (with special emphasis on a variety of environmental problems) if we are to fully comprehend the problem of achieving stated management goals within each political system.[7]

Beyond the improvements which may be implemented through domestic legislation, there is a need for new types of institutions at the international level. The rubrics of such institutions are to be found in the recommendations of the preceding chapter. In reviewing international law on pollution of lakes and rivers Bourne has suggested that private rights have given way to a system of public control and that existing:[8]

> customary international law on pollution of drainage
> basins will be largely ignored. It will be displaced
> by treaties providing for the management and control
> of international drainage basins by joint international
> agencies. Each basin will have its own regulatory
> agency, whether that agency is created directly by
> treaty or is merely the subordinate of another agency
> with a wider jurisdiction over many drainage basins.

Implicit in this suggestion is a recognition that all forms of pollution must be controlled by a single agency, and that the nature of pollution varies from region to region. New institutions to control pollution must ultimately deal with all pollution problems, although the establishment of organizations focusing on a single pollutant (such as oil) will not necessarily retard the development of a more comprehensive attack on pollution problems. The formation of such single purpose organizations may even be a logical first step, given the political realities of the 1970's. These organizations must have the necessary power to formulate management aims, to select the individuals to be restricted, and to develop the means of restriction and enforcement. Institutions must be developed for estuarine and ocean areas in addition to the more traditional river basin and lake authorities. Each of these areas will present a different spectrum of pollution problems which will have to be dealt with on a regional basis. Regional programs will have limited success in reducing or arresting the deterioration in the quality of the global environment, given the common property nature of air and water, unless such programs operate within some standardized global guidelines.

This study has examined the problem posed by the threat of international oil pollution in a specific area. An attempt has been made to assess the volition of nations to surrender sovereignty to protect the quality of the global environment. Recommendations have been made for the prevention and control of oil pollution in the study area. The prospects for positive action which would protect international environmental quality are not encouraging. Nations regard economic and defense objectives as paramount priorities. They have not been willing to surrender portions of their sovereignty because they do not feel there is an international environmental crisis. They have ignored the possibility that many resources may be permanently damaged or destroyed if preventive measures are not taken. It is paradoxical that a world learning to conquer and perhaps live in celestial space has not first addressed itself to the environmental problems of its own spaceship, the earth.

REFERENCES

1. Statement by United States Senator Edmund Muskie as quoted in
    GARDNER, R.N. "U.N. as Policeman," Saturday
    Review (August 7, 1971), p. 50.

2. SCHACHTER, O. and SERWER, D. "Maritime Pollution Problems
    and Remedies," American Journal of International Law,
    No. 65 (January 1971), p. 111.

3. These fears were confirmed in a report by Texas Instruments, Inc.
    which found substantial damage from the April 26,
    1971 oil spill at Anacortes and substantiated similar
    findings by Dr. Max Blumer on the eastern coast of the
    United States. See SCATES, S. "Grim Anacortes
    Oil Spill Report," Seattle Post-Intelligencer, October
    20, 1971, pp. A1, A12.

4. SVART, L.M. Field Burning in the Willamette Valley: A Case
    Study of Environmental Quality Control. Seattle:
    University of Washington, Department of Geography,
    unpublished M.A. Thesis (1970), p. 110.

5. ANDERSON, D. "Government and Environment: A Need for
    Public Participation," University of British Columbia
    Law Review, No. 6 (June 1971), pp. 111-114.

6. In 1969 the United States created, perhaps inadvertently, a pos-
    sible mechanism through which international implica-
    tions of domestic resource decisions might be evaluat-
    ed. The National Environmental Policy Act requires
    the federal government to undertake environmental im-
    pact studies for major resource projects involving the
    expenditure of federal funds. On the strength of this
    Act, David Anderson, a Canadian Member of Parlia-
    ment, has filed suit in a United States Court, which
    would, if successful, require the Department of the In-
    terior to evaluate the impact of Alaska oil on the Ca-
    nadian environment. A decision in favor of Anderson
    would not necessarily prohibit the building of the
    pipeline from Prudhoe Bay to Valdez, but it would es-
    tablish a useful precedent for subsequent domestic en-
    vironmental use decisions which have international im-
    plications.

7. Congress and the Environment provides such an assessment for the
         United States. Canadian scholars would do well to
         conduct a similar evaluation for the Canadian political
         system. See COOLEY, R.A. and WANDESFORDE-
         SMITH, G. (eds.) Congress and the Environment. Se-
         attle: University of Washington Press (1970).

8. BOURNE, C.B. "International Law and Pollution of Internation-
         al Rivers and Lakes," University of British Columbia
         Law Review, No. 6 (June 1971), p. 136.

POSTSCRIPT

By October 1972 concern over oil pollution in the Puget Sound
and Strait of Georgia regions had intensified, becoming even more a public issue with deep emotional overtones and international implications
than it had been in 1971. Earlier sections of this study documented the
threat of and response to oil pollution in the study region to January 1,
1972. Since that time legislation, various reports, international meetings,
and new incidents of oil spillage have altered the details, but not the substance, of the earlier analysis. These events have only confirmed the
conclusions reached in the preceding sections and added greater urgency
to the recommendations set forth in Chapter Five.

LEGISLATION SINCE JANUARY 1972

Oil Spills in British Columbia and New Legislation

In the spring of 1972 several small oil spills occurred in waters off
the British Columbia coast. These included: (1) a spill from a leaky flange
at the Standard Oil Refinery in Burnaby which resulted in a 14 mile slick,
(2) a 500 gallon discharge of Bunker C from Neptune Terminals in Vancouver harbour, and (3) a spill of 1,000 gallons of Bunker C from the Canadian destroyer, Gastineau, which polluted a half mile of beach and oy-
ster beds twelve miles north of Nanaimo. These spills were dwarfed, however, by the Panamanian registered freighter, Vanlene, which went aground
on Vancouver Island near the mouth of Barclay Sound on Tuesday, March 14,
1972, and spilled more than 37,000 gallons of heavy bunker fuel on
beaches in the area. While the Vanlene was destined for Vancouver, mariners were quick to note that: (1) the freighter was plying a route similar
to that proposed for oil shipments into Puget Sound from Alaska; and
(2) this grounding portended even more disastrous spills if the Trans-Alaska
Pipeline were constructed. They also noted, and reminded Canadians especially, that international oil pollution was as much a threat from

Canadian waters as from vessels and oil handling facilities in United
States and international waters. Conservationists also became concerned
about the inability of the federal contingency plan to meet the crisis. Oil
from the ship began to spread fast along the coastal beaches. Yet, it took
federal authorities three days to get a boom around the ship and peat moss
arrived too late because those in charge of combatting the spill forgot
that a water route would be faster than a land route. Damage from the
spill was minimized, not as a result of good contingency planning, but
rather by heavy winds and rain which, respectively, helped sweep the oil
out to sea and dissipate it.[1]

These incidents prompted the British Columbia government to introduce amendments to the Petroleum and Natural Gas Act and to the Pipelines Act which were designed to control oil spills. Under the amendments
the provincial government was given power to conscript men or equipment
to clean up oil spills and the minister was authorized to establish spill control measures and assess costs against any person responsible for the spills.
The British Columbia legislation meets none of the principles noted in
Chapter Five as necessary for the protection or compensation of victims of
transboundary pollution. It is typical of the band-aid approach to large
environmental issues, where new laws are passed to treat the effects but
not the causes of problems. These new laws do nothing to enhance the
position of British Columbia vis-a-vis an international claim for damage
and, other than specifically providing manpower, offer nothing beyond
that available in federal legislation.

Legislation in the United States

In Chapter Three proposals designed to prevent oil pollution from
tankers in United States waters were discussed. Legislation based on the
Ports and Waterways Safety Bill of 1971 has been enacted in the form of
the Navigable Waters Safety and Environmental Quality Act of 1972

(PL92-340).[2] The legislation called for prevention of accidents through navigational and traffic improvements. In addition, construction standards for tankers were tightened requiring new ships to have double hulls. Regulations for tanker operations were improved and vessels were required to have completely clean ballast discharges under the new law. Foreign vessels not meeting the standards set forth in the Act would not be able to enter United States waters after 1975. The Act makes no provisions, however, for American vessels which do not meet the proposed standards and were under construction prior to 1972. Thus, ships being constructed by Atlantic Richfield for the Alaska route would not be affected by the new law.

Pilot programs for Puget Sound passed by Congress in 1971 were expanded by the Act largely through the efforts of Senator Warren Magnusson. The United States Coast Guard has continued development of a traffic separation system with buoy-marked lanes for shipping inside Dungeness Light extending south to Tacoma and north to Cherry Point. A shipping-communication center is under construction at Pier 90 in Seattle which will have voice contact with skippers or pilots on all vessels in the Sound. Plans call for the Coast Guard to award contracts for production of an oil-recovery system that could be delivered by air to a stranded ship to pump oil from its tanks, for delivery of a high seas containment boom, and for development of mechanical skimmers. In addition, an experimental Coast Guard program which had personnel at more than 90 percent of oil transfers exceeding 5,000 barrels between tankers, barges, and shore installations along Puget Sound is being taken over by the State of Washington as part of its monitoring system.[3] This monitoring system will include inspection of oil refineries and handling functions as well as oil transfers. All of these actions enhance the ability of the United States to prevent oil pollution in her own waters, but, of course, offer no comprehensive solution to the larger oil pollution threats from Canada and international waters.

THE EXISTING THREAT

Coastal Traffic

Meanwhile, the existing threat of international oil pollution has increased with the expansion of coastal traffic and approval of the Alaska Pipeline. In addition to the small spills noted earlier in this chapter, oil barges and tankers continue to run aground although all do not result in oil discharges. Tankers have begun to supply some of the needs of the Cherry Point Refinery. Oil continues to be transported in large quantities through Puget Sound and the Strait of Georgia. In response to criticisms from environmentalists Alyeska, the pipeline service company, has quoted detailed figures on increases that can be expected in oil tanker traffic if the Alaska Pipeline is built. North Slope oil will add about 70 tanker calls per year in Puget Sound as soon as the Pipeline begins operation. This would go up to about 80 additional calls annually when the Pipeline reaches full capacity seven years later. The pipeline company notes that some oil will be moved by foreign vessels and will, therefore, because of the Jones Act, move to countries other than the United States. If the company's estimates are correct, tankers from Alaska would constitute only about 1 percent of the total ship traffic in Puget Sound by 1980.[4] Still, rumours of an oil pipeline to the Midwest, with terminal facilities in Puget Sound, continue to persist. In the event of this development all estimates quoted by the oil pipeline company would be meaningless. The essential point is that no matter what estimates might be quoted, the oil companies are under no obligation to control either the volume of oil or number of vessels engaged in oil trade within Puget Sound or the Strait of Georgia. Under these conditions and the existing trade patterns discussed in Chapter Three, the threat of oil pollution can do nothing but increase.

217

The Alaska Pipeline

Construction of the Alaska Pipeline moved much closer to
reality in 1972. In 1971 the United States Department of the Interior released the draft environmental impact statement on the Alaska Pipeline.
Hearings were held, criticisms were voiced and the Department began
preparation of a final report for publication in 1972. On March 20, 1972
the Interior Department released the report which documented the environmental impact. The proposed Alaska Pipeline appeared to be the worst
choice in terms of environmental damage. Nevertheless, Interior Secretary Rogers Morton announced his intention to approve the Alyeska application and issue a permit for construction. The reaction was loud and
swift. Twenty-three Senators requested public hearings on the final statement. Other groups and individuals, the Wilderness Society, Friends of
the Earth, the Environmental Defense Fund, the Canadian Wildlife Federation, and David Anderson, Member of the Canadian Parliament, antic-
ipating Morton's actions, had earlier filed an injunction to require the
Interior Department to hold hearings. Representative Les Aspin introduced
resolutions that would hold up construction until a comprehensive study of
the proposed pipeline through Canada had been completed. Further,
congressional assent would be required prior to building any pipeline
through Alaska. Rogers Morton countered that he would be willing to
entertain any concrete proposals for an oil pipeline through Canada, but
none have been presented.[5]

On August 15, 1972 Judge George Hart denied injunctions against
the pipeline, ruled that the environmental impact statement was adequate,
and announced that permits for construction could be issued. Almost immediately plans were announced to appeal Judge Hart's decision to the
Supreme Court. As a result further delays of 12 to 18 months can be expected before any construction could begin on the pipeline.

The court case and Morton's decision to approve the pipeline have raised two interesting and important points. The interventions of the Canadian Wildlife Federation and David Anderson are an important precedent. This marks the first time foreign nationals have been allowed to intervene in an environmental issue before a United States court. Such action may prove to be significant in future environmental conflicts between Canada and the United States which have transboundary implications. Morton's decision demonstrates that the National Environmental Policy Act of 1969 may not have the impact that those espousing environmental protection once envisioned. It would appear that statements could be issued noting severe environmental damage, but projects would be approved on the grounds that other considerations such as national security and economics had a higher priority in the decision. Whatever the impact these developments may have on environmental litigation, it is clear that the Alaska Pipeline has moved much closer to construction and that this pipeline will increase the probability of a major oil spill in Puget Sound and the Strait of Georgia.

The Vagners and Mar Report

Three studies published in 1972 document in great detail the range of the threat and the magnitude of possible impacts if a major oil spill were to occur. These include studies conducted by Juris Vagners and Paul Mar, Howard Paish and Associates, and the Master Mariners of Canada.[6] Vagners and Mar have prepared a comprehensive analysis of oil on Puget Sound. While the study offers different predictions for chances of a major oil spill from those of Honeywell quoted in Chapter Three, the prognostications are just as gloomy. Vagners and Mar estimate that given the worst conditions there could be up to nine ship collisions per year by 1990 with the likelihood that one of the vessels involved will be an oil tanker. At the other end of the probability scale, the lowest estimates of a tanker being gouged open in a collision puts the risk at one in one

219

hundred for the hazard period to 1990. In terms of total volume of crude oil, the study predicts 20.2 million tons of crude oil will be carried through the area by 1980, although this figure does not account for crude oil shipped through Canadian waters.

Predictions of collisions and total volume can vary greatly depending on the assumptions made about growth of demand, total volume of crude and refined oil, and a host of other variables. The most valuable part of this study lies not in the predictions, but rather in the assessment of the level of preparedness for dealing with oil spills and the preventive procedures to see that they do not happen in the first place. The report is extremely critical of the oil industry. Estimates place industry expenditures for oil spill prevention at 1.4 million dollars. Vagners and Mar charge that at this rate of outlay the industry is guilty of short changing the Northwest. These charges are documented fully. For example, in spite of the prevalence of strong winds in the area only one of the state's four major refinery docks is sheltered. Transfer facilities, which have been the source of most major oil spills to date, lack modern equipment and are rated from "fair to not so good" compared to the most modern facilities. Using the same criteria, the Atlantic Richfield prevention facilities, the newest in the area, were rated as fair, although better than those at other locations including U.S. Navy facilities. The report notes that the Coast Guard, the Environmental Protection Agency, and the Washington State Department of Ecology have limited numbers of trained manpower available and lack specialized equipment to deal with spills. This lack of manpower and equipment has resulted in poor cleanup performances in the spills that have already occurred. For instance, in the Anacortes spill discussed earlier in Chapter Three, only 450 gallons were recovered from the total spill of 242,000 gallons. The response capability of contingency planning is regarded as minimal at best, but largely unknown. In addition, there is jurisdictional confusion between agencies enforcing federal and state laws, a lack of funds to back up existing laws at the state level, and a failure to adapt national prevention and control programs to local conditions.[7]

The Paish Report

The Paish report contains a comprehensive analysis of the effect of oil pollution in Canada. It outlines the problem of oil pollution at a general level and analyzes the impact of a number of simulated spills. Most of the failures noted in the Vagners and Mar study were found to be applicable to the Canadian case. Recommendations in the Paish report note the need for more research and broadly mirror the proposals advanced in Chapter Five,[8] but are more detailed with respect to unilateral actions that Canadian authorities could undertake to enhance their response capability.

The Master Mariners of Canada Report

The report prepared by the Master Mariners of Canada makes explicit recommendations for the British Columbia Coast. The Master Mariners conclude, as does the Paish report, that British Columbia is ill prepared to deal with major oil spills from tankers. They recommend a multi-million dollar investment in prevention, new laws requiring cellular construction for ships frequenting Canadian waters, large salvage stations with hovercraft that can respond quickly to spills, and a 100 million dollar floating drydock for making major repairs.

In all three reports there is explicit recognition of the international problems that can result from oil pollution in the area. Two, the Vagners and Mar and Master Mariners of Canada reports, call for joint arrangements in prevention and cleanup operations. Only the Paish report, however, takes cognizance of most of the basic issues involved in removing all externalities created by spilled oil. It is apparent from all of these reports that each jurisdiction involved in the region - Canada, British Columbia, the United States, Washington, and various counties and

221

municipalities – has its own particular approach to oil pollution prevention, control, and damage compensation, which has resulted in confusion and inefficiency in dealing with individual spills. Against this background and the reaction to the Cherry Point spill described below, the proposal for an international commission advanced in Chapter Five becomes even more credible and urgent.

THE CHERRY POINT OIL SPILL

On Sunday, June 4, 1972 at 6 a.m. a spill from the Liberian registered tanker World Bond occurred at the Cherry Point Refinery of Atlantic Richfield (ARCO), built specifically to utilize Alaskan oil. This spill, estimated at between 4,000 and 12,000 gallons, had serious international implications. Oil from the spill quickly spread out into the Strait of Georgia and damaged adjacent beaches near Tongue Point, Birch Point, Birch Bay, and Point Whitehorn in the United States. Attempts were made to clean up the oil and Atlantic Richfield gave assurances that the oil would probably not affect Canadian waters.

Reaction to the spill was predictable. Barry Mather, the Member of Parliament for Surrey, asked Prime Minister Trudeau to inform President Nixon that what happened is just a small example of what will occur if the shipment of oil in tankers down the Pacific Coast from Alaska goes ahead. Preservationist groups such as the Society for Pollution and Environmental Control (SPEC), the Canadian Wildlife Federation, and the Sierra Club joined in renewed protests against the Alaska Pipeline.

The worst fears of these groups were confirmed on Monday night, June 5, 1972 when oil hit Canadian shores at Crescent Beach. The apparent confusion when the oil impregnated the beaches confirmed earlier criticism of contingency planning and international cooperation. ARCO had been monitoring the spill from the air but neglected to warn Canadian officials that the oil was approaching land in British Columbia. ARCO sent officials into Canada to help in cleanup operations but neither they nor the officials representing the federal government had much experience in cleaning up oil spills. Meanwhile, the oil continued to spread and by Thursday had washed ashore further north at Whiterock. Reports of dead fish, oil soaked birds, and damage to beach facilities, so common in all spills, were received by ARCO. The company maintained, however, that this was a minor spill and there was little damage.

Political reaction at the provincial, state, and national levels was initially a reflection of past policy. The House of Commons in Ottawa rejected a motion to invite Rogers Morton to view the spill, but later passed motions protesting the spill and urged referral of the problem to the International Joint Commission. The British Columbia government initially maintained that it was not involved in oil spill problems because navigation is a federal responsibility under the constitution, but subsequently, requested that Washington State agree to a joint study of the oil pollution problem in the area. Washington State moved under state law to prosecute those responsible for the spill.

Only at the municipal level was there much decisive action. The municipality moved quickly to combat the spill and secured a promise from ARCO to pay for all cleanup costs and damages. It later began, with the help of federal authorities, to investigate the possibility of suing ARCO through registered judgment. Under a registered judgment the jurisdiction of a Canadian court would be valid because the cause of action arose in Canada even though the defendant was not present. The problem, however, is whether such a judgement could be registered in a United States Court without a bitter fight. A registered judgement, like the Canadian intervention in the Alaska Pipeline court case, is a unique application of the law to an environmental issue. It could be an important instrument in reducing the impact of transboundary pollution and enhancing efforts to improve environmental quality.

Obviously, the municipality preferred a negotiated settlement and when ARCO agreed, the court action became a mute point. Still, if ARCO refuses to meet all the costs of the June 4, 1972 spill, or any further spills, then registered judgements might serve to help victims in the absence of international compensation procedures as outlined in Chapter Five.

RESPONSE TO THE CHERRY POINT SPILL

The Cherry Point Spill and the Stockholm Conference

As an isolated incident the Cherry Point oil spill would have been a minor irritant between Canada and the United States but, as noted earlier in this study, oil has become a major issue in Canadian-United States relations. Canadian opposition to the proposed oil tanker route through the Arctic and down the Pacific Coast of North America triggered the Arctic Waters Pollution Prevention Act, amendments to the Canada Shipping Act, and Canadian efforts to overthrow existing international concepts concerning freedom of the seas. These efforts will culminate in 1973 when a new Law of the Sea Conference will be convened. Canada, however, took major steps to control international pollution at the Stockholm Conference on the Human Environment in June. The Conference was in session and considering efforts to control pollution when the spill at Cherry Point occurred.

Canada and several of the developing nations directly challenged the world's major shipping nations, particularly the United States, Britain, Japan, Greece, and Liberia, over the issue of bringing ocean pollution under international law. Twenty-six principles were endorsed by the Stockholm Conference and forwarded to the 1973 Law of the Sea Conference. The propositions supported the notion that the principle of freedom of the seas had been unfortunately interpreted as including the freedom to overfish and pollute. The principles would formally introduce into

224

international law the notions that nations have a responsibility not to pollute, that nations may exercise special authority outside territorial waters
to protect their shores, and that individual nations may take action to
benefit mankind as a whole. These principles would, for instance, allow
a nation to legally firebomb an oil tanker in danger of breaking up in
international waters and fouling that nation's coasts, as well as to manage
a fishery resource in international waters.

It would be false to suggest that the principles were endorsed by
the Conference because of the Cherry Point oil spill. However, there is
no doubt that the timing of the spill and the Stockholm Conference strengthened Canadian arguments for international pollution control, contributed
to ARCO's decision to pay for all costs associated with the spill, and
caused the United States to reconsider its opposition to study the threat of
oil pollution along the international border.

Canadian and United States Responses to the Cherry Point Spill

Prior to the Cherry Point spill the Canadian government was com-
placently preparing for talks with United States officials over the oil tanker route and for negotiations with British Columbia over control of oil
tanker shipments in Canadian waters off the Pacific Coast. The spill, however, prompted more decisive action. Jack Davis, Canadian Minister of
the Environment, noted the intention of the government to settle claims
through the Maritime Pollution Claims Fund described in Chapter Three
and to recoup for any damage, even if that means the possibility of a law
suit in United States courts. The government also formally asked the
United States for prompt payment of damages and cleanup costs, sought
permission, as noted earlier, from the United States to request the IJC to
investigate the oil tanker threat, and attempted to have the United States
jointly designate Puget Sound and the Strait of Georgia as an international
water quality management area. The Canadian public, however, was more

thoroughly aroused. An editorial in the <u>Vancouver Province</u> reflected
their feelings:[9]

> It's not sufficient for ARCO to promise to pay for all
> the cleanup costs. It should offer a full and honest
> explanation of what went wrong so that Canada can
> have at least some assurance the same mistakes won't
> happen again. Nor is compensation for cleanup costs
> sufficient recompense for the damage to Canadian
> beaches. Canada may be entitled to punitive damages
> against ARCO.

While much venom was directed at the United States and ARCO,
the public also became more concerned about the ability of the federal
contingency plan to deal with spills, however small. The government
was urged to give some substance to the plan even if that meant considerable expenditures to acquire equipment and train manpower.

Reactions in the United States were mixed. Concerned about the
impact of the Cherry Point spill on the Stockholm Conference, the United
States sought to diffuse the problem and play down the impact of the spill.
While the United States acknowledged the need for prompt payment for
damage and cleanup costs, it rejected all other proposals from Canada.
A representation from Senators Jackson and Magnusson and Representative
Meeds of Washington State for a Canada-United States Oil Spill Conference to discuss coordinated actions against oil spills was ignored. Instead,
the United States suggested formation of a United States-Canada transboundary environmental committee which would study all environmental
problems along the border.

Governor Evans of Washington State agreed to a study of the oil
pollution problem in Puget Sound and the Strait of Georgia. He also
noted his intention to re-introduce legislation requiring a prevention and
monitoring program to be funded by the oil industry, even though the legislature had failed to act earlier in the year. The control program would
consist of: (1) continuation of oil spill investigation and cleanup programs,

(2) new regulations for handling oil, (3) monitoring of oil transfers, and (4) an inspection and an inventory of all water-oriented oil-handling facilities.

The Aftermath

In July 1972 British Columbia and Washington agreed, through a memorandum signed at the Blaine Peace Arch, to work toward a joint monitoring and inspection program, to plans of actions to cope with oil spills, and exchanges of information and mutual aid. British Columbia has made plans to tap the Provincial Disaster Fund to fund the program. Washington State placed the memorandum within the broader plan to control oil pollution discussed above.

Later, in July, Jack Davis met with Russell Train, Chairman of the Council on Environmental Quality, to discuss transboundary environmental problems. The two officials affirmed each nation's support of the principle that each state is responsible for the damage its pollution causes to another and agreed to work toward joint contingency planning for oil pollution on both the Pacific and Atlantic Coasts. United States officials, however, postponed serious discussion of management area proposals because of concern over legal implications.

While it is true that oil pollution in the region has been recognized as more of an international problem in 1972 than at any previous time, it is also true that the advances which have been made are cosmetic solutions to the real problem. Contingency plans remain paper creations, laws continue to be focused on prohibiting pollution and establishing sanctions, and there is still no system to internalize the full range of costs associated with oil spillage. Improvements that have been fashioned help to correct some of the flaws in domestic responses to the threat, but that threat continues to grow. The international problem has been recognized and is now being studied, and local authorities, up to the subnational level, have taken positive action, but some national governments have been reluctant to move toward a new order for international pollution prevention and control. It remains to be seen if the Stockholm Conference and the 1973 Law of the Sea Conference will help achieve that new order and whether institutions reflecting the new order, such as that suggested in Chapter Five, will be able to internalize the externalities associated with pollution problems such as oil spillage.

REFERENCES

1. Many of the details of this section are drawn from the Seattle Post Intelligencer, Seattle Times, Vancouver Province, and Vancouver Sun which have given extensive coverage to the oil pollution problem along the Pacific Coast.

2. United States, Congress. Navigable Waters Safety and Environmental Quality Act of 1972, Public Law 92-340, 92nd Congress, 2nd session, H.R. 8140. Approved July 10, 1972.

3. FLAJSER, S. and WENK, E., Jr. "The Impact of Alaskan Oil Transport on the Marine West Coast," paper presented at the ASCE and ASME National Transportation Engineering Meeting, Seattle, Washington, July 30, 1971.

4. WILLIAMS, H. "Puget Sound Important Link in Oil Delivery," Seattle Times, May 14, 1972, p. A26; BROWN, R. C. "The Proposed Trans Alaska Pipeline System: Potential Highway to the North American Arctic," Professional Geographer, No. 23 (January 1971), pp. 15-18; HAYNES, J. B. "North Slope Oil: Physical and Political Problems," Professional Geographer, No. 24 (February 1972), pp. 17-22.

5. ZELNICK, C. R. "The Darkness at the End of the Pipeline," The Living Wilderness, No. 36 (Summer 1972), pp. 6-12.

6. VAGNERS, J., Director and MAR, P., Coordinator. Oil on Puget Sound - An Interdisciplinary Study in Systems Engineering. Seattle and London: University of Washington Press (1972); Howard Paish and Associates. The West Coast Oil Threat in Perspective, A report prepared for Environment Canada. Vancouver (April 1972); Company of Master Mariners of Canada. A Report of the Potential Menace of Large Oil Tankers Operating between Alaska and Cherry Point. Vancouver (April 1972).

7. VAGNERS and MAR, op. cit., pp. 349-359.

8.   An earlier complete draft of this manuscript was made available
       to Howard Paish and Associates of Vancouver when
       the author served as a sub-consultant to the study.

9.   "Remote Oil Spill on the Doorstep," Vancouver Province,
       June 8, 1972, p. 4.

SELECT BIBLIOGRAPHY

Manuscript Sources

Theses

Langran, K.J. The Political and Administrative Control of Water Pollution in International River Basins. Seattle: University of Washington, Department of Geography, unpublished M.A. Thesis
(1968).

Svart, L.M. Field Burning in the Willamette Valley: A Case Study of
Environmental Quality Control. Seattle: University of Washington, Department of Geography, unpublished M.A. Thesis (1970).

Wandesforde-Smith, G. A Comparative Analysis of American and Cana-
dian Governmental Arrangements for the Development of Regional
Water Policy in the Columbia River Basin. Seattle: University of
Washington, Department of Political Science, unpublished Ph.D.
Dissertation (1971).

Miscellaneous Unpublished Manuscript Sources

Benjamin, K.C. "Water Pollution Can Be Most Efficiently Controlled
Via a System of Effluent Charges," paper presented in Planning
506, University of British Columbia (1969), typewritten.

Clingan, T.A., Jr. and Springer, R. "International Regulation of Oil
Pollution," working paper for the International Law Panel of the
President's Commission on Marine Science, Engineering, and Re-
sources (undated), typewritten.

Cunningham, P. "Comment on the Washington Oil Spill Act of 1970,"
forthcoming paper for the University of Washington Law Review
(1971), typewritten.

Davis, J. "Marine Parks for More People," address to the Save Our
Parkland Association, Vancouver (November 6, 1970), mimeo.

Dunn, J.D. "Oil Pollution of the Sea," paper presented to the Oil and
Gas Seminar, Faculty of Law, University of British Columbia
(April 30, 1970), typewritten.

Glude, J.B. "Information Requirements for Rational Decision Making in
the Control of Coastal and Estuarine Pollution," paper given to
the FAO Technical Conference on Marine Pollution and its Effects

on Living Resources and Fishing, Rome, Italy (December 9-18, 1970), mimeo.

Honeywell, Marine Systems Center. "A Proposed Automated Marine Traffic Advisory System for Puget Sound," Seattle (November 6, 1970), mimeo.

International Convention on the Establishment of an International Fund for Compensation for Oil Pollution Damage (December 18, 1971), manuscript copy, 42pp.

Legault, L.H.J. "The Freedom of the Sea: A License to Pollute?," paper presented at the Symposium on International Legal Problems of Pollution under the auspices of the Canadian Branch of the International Law Association, sponsored by the University of British Columbia and the Department of External Affairs, Ottawa and held in Vancouver (September 8-11, 1970), mimeo.

Oceanographic Institute of Washington, Subcommittee on Oil Spill Taxation. Minutes of Meeting (December 9, 1970), typewritten.

"Proposed Coastal Waters Protection Act of 1971, State of Washington" (undated), manuscript copy, mimeo, 13pp.

Rodgers, W.H., Jr. "Open Letter to Members of the Legislature," letter to the Washington State Legislature (January 7, 1971), mimeo.

Transmountain Oil Pipe Line Company. "General Article," Vancouver (1969), supplemented (1970), mimeo.

Yoshioka, T.T. "Problems of International Control of Oil Pollution of the Sea," paper presented to Professor Rodgers' law seminar, University of Washington (undated), typewritten.

Printed Sources

Government

1. United States

Department of Defense, Department of the Army, Corps of Engineers. Waterborne Commerce of the United States, Calendar Year, 1968. Washington: United States Government Printing Office (1970).

Departments of Interior and Transportation. Oil Pollution - A Report to the President. Washington: United States Government Printing Office (1968).

Department of State. Department of State Bulletin. Washington: United States Government Printing Office (1950-1971).

Department of Transportation, United States Coast Guard. "Maritime Environmental Protection Activities of the Coast Guard," Commandant Notice 3010, August 20, 1970. Washington: United States Coast Guard (1970).

Department of Transportation, United States Coast Guard. Seattle Coastal Region Oil and Hazardous Materials Pollution Contingency Plan. Seattle: Thirteenth Coast Guard District (December 1, 1970).

Executive Office of the President, Office of Science and Technology. Offshore Mineral Resources, A Challenge and an Opportunity, Second Report of the President's Panel on Oil Spills. Washington: United States Government Printing Office (1969).

Executive Office of the President, Office of Science and Technology. The Oil Spill Problem, First Report of the President's Panel on Oil Spills. Washington: United States Government Printing Office (1968).

House of Representatives, Congressional Delegation to the Third Extraordinary Session of the Intergovernmental Maritime Consultative Organization. Report on International Control of Oil Pollution, Document 628, 90th Congress, 1st Session. Washington: United States Government Printing Office (1967).

House of Representatives, Committee on Merchant Marine and Fisheries. Hearings on Oil Pollution, Serial 91-4. Washington: United States Government Printing Office (1967).

House of Representatives, Committee on Public Works. Laws of the United States Relating to Water Pollution Control and Environmental Quality, Committee Print 91-33, 91st Congress, 2nd Session. Washington: United States Government Printing Office (1970).

House of Representatives, Committee on Public Works. Hearings on Federal Water Pollution Control Act Amendments, Serial 91-1. Washington: United States Government Printing Office (1969).

Senate, Committee on Commerce. Environmental Activities of International Organizations. Washington: United States Government Printing Office (1971).

Senate, Committee on Interior and Insular Affairs. Congress and the Nation's Environment — Environmental Affairs of the 91st Congress. Washington: United States Government Printing Office (1971).

Senate, Committee on Public Works. Oil Pollution of the Marine Environment — A Legal Bibliography. Washington: United States Government Printing Office (1971).

Senate, Committee on Public Works, Subcommittee on Air and Water Pollution. Hearings on the Proposed Regulations of the Department of the Interior on Oil Pollution Under the Water Quality Improvement Act of 1970. Washington: United States Government Printing Office (1970).

2. Washington State

Laws of Washington, 2nd Extraordinary Session, 1971, Chapter 180. Olympia: State of Washington (1971).

Revised Code of Washington. Olympia: State of Washington (1970).

Department of Ecology. Oil Spill Action Plan. Olympia: State of Washington (1972).

Department of Ecology. Water Pollution Control Laws. Olympia: State of Washington (1970).

Department of Ecology. Laws and Oil Spill Emergency Procedures. Olympia: State of Washington (1970).

Department of Ecology. "Reported Oil and Hazardous Material Spills in Washington State," monthly since January 1970. Olympia: State of Washington (1970).

3. Canada

Department of Transport, Marine Operations. Interim Federal Contingency Plan for Combatting Oil and Toxic Material Spills. Ottawa: Department of Transport (1970), mimeo.

National Energy Board. Energy Supply and Demand in Canada and Ex-

port Demand for Canadian Energy, 1966-1990. Ottawa: Queen's
Printer (1969).

Parliament, House of Commons. Official Report of Debates. (1867–
1972), various issues.

Parliament, House of Commons. Order 209. (November 12, 1971);
Aide-Memoire to U.S.A. Government (August 18, 1971) and
attachments.

Parliament, House of Commons, Special Committee on Environmental
Pollution. Minutes of Proceedings and Evidence. Ottawa: Information Canada and Queen's Printer (1970-1971).

Parliament. Statutes. (Sessional issues since 1967).

Science Council of Canada. Canada, Science and the Oceans, Report
Number 10. Ottawa: Science Council of Canada and Information Canada (1970).

McTaggart-Cowan, P.D., Sheffer, H., and Martin, M.A. Report of
the Task Force — Operation Oil. Ottawa: Ministry of Transport
and Information Canada (1970 and 1971), 3 vols.

4. British Columbia

Legislative Assembly. Statutes. (Sessional issues since 1900).

5. Other

State of Michigan. Annual Statutes (1960-1971).

University of Maine, School of Law, United States National Science
Foundation, Office of Sea Grant Programs. Maine Law Affecting Marine Resources. University of Maine, School of Law
(1969), 4 vols.

Journal Articles

Alexander, L.M. "National Jurisdiction and the Use of the Sea," Natural Resources Journal, vol. 8 (July 1968), pp. 373-400.

Avins, A. "Absolute Liability for Spillage," Brooklyn Law Review, vol. 36 (Spring 1970), pp. 359-367.

Baldwin, M.F. "Public Policy on Oil — An Ecological Perspective," Ecology Law Quarterly, vol. 1 (Spring 1971), pp. 245-303.

Bator, F.M. "The Anatomy of Market Failure," Quarterly Journal of Economics, vol. 72 (August 1958), pp. 351-379.

Bellamy, D.J. "Effects of Pollution from the Torrey Canyon on Littoral and Sub-littoral Ecosystems," Nature, vol. 216 (December 23, 1967), pp. 1170-1173.

Bilder, R.B. "Canadian Arctic Waters Pollution Prevention Act: New Stresses on the Law of the Sea," Michigan Law Review, vol. 69 (November 1970), pp. 1-54.

Bone, Q. and Holme, N. "Lessons from the Torrey Canyon," New Scientist, vol. 39 (September 5, 1968), pp. 492-493.

Bone Q. and Holme, N. "Oil Pollution — Another Point of View," New Scientist, vol. 37 (February 15, 1968), pp. 365-366.

Boyle, C.L. "Oil Pollution of the Sea: Is the End in Sight?," Biological Conservation, vol. 1 (July 1969), pp. 319-327.

Brown, G., Jr. and Mar, B. "Dynamic Economic Efficiency of Water Quality Standards or Charges," Water Resources Research, vol. 4 (December 1968), pp. 1153-1159.

Buchanan, J.M. and Stubblebine, W.C. "Externality," Economica, vol. 29 (November 1962), pp. 371-384.

Busch, D.D. and Mears, E.L. "Ocean Pollution: An Examination of the Problem and An Appeal for International Cooperation," San Diego Law Review, vol. 7 (July 1970), pp. 574-604.

Byrne, J. "Canada and the Legal Status of Ocean Space in the Canadian Arctic Archipelago," University of Toronto Faculty of Law Review, vol. 28 (1970), pp. 1-16.

Caplan, N. "Offshore Mineral Rights: anatomy of a federal-provincial conflict," Journal of Canadian Studies, vol. 5 (February 1970), pp. 50-61.

Castle, E.N. and Stoevener, H.H. "Water Resources Allocation, Extramarket Values, and Market Criteria: A Suggested Approach," Natural Resources Journal, vol. 10 (July 1970), pp. 532-544.

Clagett, B.M. "Survey of Agreements Providing for Third Party Resolution of International Water Disputes," American Journal of International Law, vol. 55 (July 1961), pp. 645-669.

Clark, R.B. "Organization Against Oil," New Scientist, vol. 43 (September 25, 1969), pp. 658-660.

Christy, F.T., Jr. "Efficiency in the Use of Marine Resources," Resources for the Future, Reprint No. 49 (1964), 8pp.

Coase, R.H. "The Problem of Social Cost," Journal of Law and Economics, vol. 3 (October 1960), pp. 1-44.

Crowe, B.L. "The Tragedy of the Commons Revisited," Science, vol. 166 (November 28, 1969), pp. 1103-1107.

Crutchfield, J.A. "An Economic Evaluation of Alternative Methods of Fishery Regulation," Journal of Law and Economics, vol. 4 (October 1961), pp. 131-143.

Crutchfield, J.A. "Common Property Resources and Factor Allocation," Canadian Journal of Economics and Political Science, vol. 22 (August 1956), pp. 292-300.

Crutchfield, J.A. "Valuation of Fishery Resources," Land Economics, vol. 28 (May 1962), pp. 145-154.

Davis, O.A. and Whinston, A. "Externalities, Welfare, and the Theory of Games," Journal of Political Economy, vol. 70 (June 1962), pp. 241-262.

Dean, C.J. "Markets for Alaskan Oil," Scottish Geographical Journal, vol. 87 (September 1971), pp. 147-150.

Dobbert, J.P. "Water Pollution and International River Law," Yearbook of the Association of Attenders and Alumni of the Hague Academy of International Law, vol. 35 (1965), pp. 60-99.

Ellis, H. and Fellner, W. "External Economies and Diseconomies," American Economic Review, vol. 33 (September 1943), pp. 493-511.

Fay, J.A. "Oil Spills: The Need for Law and Science," Technology Review, vol. 72 (January 1970), pp. 32-35.

Finnie, R.S. "North American Arctic Petroleum Development," Canadian Geographical Journal, vol. 83 (November 1971), pp. 146-161.

Fox, W.T.R. "Science, Technology, and International Politics," International Studies Quarterly, vol. 12 (March 1968), pp. 1-15.

Gabrielson, I.N. "Oil Pollution," National Parks Magazine, vol. 44 (March 1970), pp. 4-9.

Gagné, M. "The Legal Status of the Waters Met by the Manhattan During her Voyage Through the Arctic," Les Cahiers de Droit, vol. 11 (April 1970), pp. 66-73.

Galway, M. "What We Can Win in the Arctic," Saturday Night, vol. 85 (April 1970), pp. 23-25.

Goldie, D.M.M. "Effect of Existing Uses on the Equitable Apportionment of International Rivers: A Canadian View," University of British Columbia Law Review, vol. 1 (April 1963), pp. 763-776.

Gordon, H.S. "An Economic Approach to the Optimum Utilization of Fishery Resources," Journal of the Fisheries Research Board of Canada, vol. 10 (October 31, 1953), pp. 442-457.

Hardin, G. "Finding Lemonade in Santa Barbara Oil," Saturday Review, vol. 52 (May 10, 1969), pp. 18-21.

Hardin, G. "The Tragedy of the Commons," Science, vol. 162 (December 13, 1968), pp. 1243-1248.

Harris, M.B. and Lovett, A. "Recent Developments in the Law of the Sea: A Synopsis," San Diego Law Review, vol. 7 (July 1970), pp. 627-673.

Head, I.L. "The Canadian Offshore Minerals Reference: The Application of International Law to a Federal Constitution," University of Toronto Law Journal, vol. 18 (1968), pp. 131-157.

Healy, N.J. "The C.M.I. and IMCO Draft Conventions on Civil Liability for Oil Pollution Damage," Journal of Maritime Law and

Commerce, vol. 1 (October 1969), pp. 93-105.

Healy, N.J. "The International Convention on Civil Liability for Oil
Pollution Damage, 1969," Journal of Maritime Law and Commerce, vol. 1 (January 1970), pp. 317-323.

Healy, N.J. and Paulsen, G.W. "Marine Oil Pollution and the Water
Quality Improvement Act of 1970," Journal of Maritime Law and
Commerce, vol. 1 (July 1970), pp. 537-572.

Heeney, A. "Along the Common Frontier – the International Joint Commission," Behind the Headlines. Toronto: Canadian Institute of
International Affairs (July 1967), pp. 2-18.

Henkin, L. "Arctic Anti-Pollution: Does Canada Make-or-Break International Law?," American Journal of International Law, vol. 65
(January 1971), pp. 131-136.

Holcomb, R.W. "Oil in the Ecosystem," Science, vol. 166 (October
10, 1969), pp. 204-206.

Hovanesian, A., Jr. "Post Torrey Canyon: Toward a New Solution to
the Problem of Traumatic Oil Spillage," Connecticut Law Review,
vol. 2 (Spring 1970), pp. 632-647.

Jennings, R.Y. "The Limits of Continental Shelf Jurisdictions: Some
Possible Implications of the North Sea Case Judgement," International and Comparative Law Quarterly, vol. 18 (October 1969),
pp. 819-832.

Johnson, R.W. "Effect of Existing Uses on the Equitable Apportionment
of International Rivers: An American View," University of British
Columbia Law Review, vol. 1 (December 1960), pp. 389-398.

Jordan, F.J.E. "Recent Developments in International Environmental
Pollution Control," McGill Law Journal, vol. 15 (1969), pp.
279-301.

Kneese, A.V. and d'Arge, R.C. "Pervasive External Costs and the Response of Society," Resources for the Future, Reprint No. 80
(July 1969), pp. 87-115.

Konan, R.W. "The Manhattan's Arctic Conquest and Canada's Response
in Legal Diplomacy," Cornell International Law Journal, vol. 3
(Spring 1970), pp. 189-204.

Krueger, R.B. "International and National Regulation of Pollution from Offshore Oil Production," San Diego Law Review, vol. 7 (July 1970), pp. 541–573.

Kuhn, A.K. "The Trail Smelter Arbitration — United States and Canada," American Journal of International Law, vol. 32 (October 1938), pp. 785–788.

Landis, H. "Legal Controls of Pollution in the Great Lakes Basin," Canadian Bar Review, vol. 38 (January 1971), pp. 49–62.

Lehr, E.E. "Coast Guard Fights Pollution at Sea," Data on Civil and Defense Systems, vol. 15 (January 1970), pp. 22–25.

Lester, A.P. "River Pollution in International Law," American Journal of International Law, vol. 57 (October 1963), pp. 828–853.

Little, C.H. "Giant Bulk Carriers," Canadian Geographical Journal, vol. 77 (December 1968), pp. 108–115.

Little, C.H. "Offshore Exploration for Gas and Oil," Canadian Geographical Journal, vol. 77 (October 1968), pp. 108–115.

Livingston, D. "Pollution Control: An International Perspective," Scientist and Citizen, vol. 10 (September 1968), pp. 172–182.

Loftas, T. "Decade for Ocean Escalation," New Scientist, vol. 47 (July 2, 1970), pp. 31–33.

Logan, R. "Mineral Resource Administration Lines," Professional Geographer, vol. 23 (April 1971), pp. 160–163.

Lohne, A.A. "Oil Pollution of Coastal and Inland Waters of the United States Under the Water Quality Improvement Act of 1970," Insurance Counsel Journal, vol. 38 (January 1971), pp. 49–62.

Lucas, A.R. "Water Pollution Control Law in British Columbia," University of British Columbia Law Review, vol. 4 (May 1969), pp. 56–86.

McCaull, J. "The Black Tide," Environment, vol. 11 (November 1969), pp. 2–16.

McCoy, F.T. "Oil Spill and Pollution Control: The Conflict Between State and Maritime Law," George Washington Law Review, vol.

40 (October 1971), pp. 97-122.

McDonald, J. "Oil and the Environment: The View From Maine," Fortune, vol. 83 (April 1971), pp. 84-89, 146-147, 150.

McDougall, I.A. "The Development of International Law with Respect
to Trans-Boundary Water Resources: Co-operation for Mutual Advantage or Continentalism's Thin Edge of the Wedge?," Osgoode
Hall Law Journal, vol. 9 (November 1971), pp. 261-311.

Mackenzie, K.D. "Interprovincial Rivers in Canada: A Constitutional
Challenge," University of British Columbia Law Review, vol. 1
(September 1961), pp. 499-512.

Maloney, F.E. and Ausness, R.C. "Water Quality Control: A Modern
Approach to State Regulation," Albany Law Review, vol. 35
(Fall 1970), pp. 28-59.

Mapes, G. "Troubled Waters," Wall Street Journal, vol. 174 (November 26, 1969), pp. 1, 20.

Maywhort, W.M. "International Law — Oil Spills and Their Legal Ramifications," North Carolina Law Review, vol. 49 (August 1971),
pp. 996-1003.

Meade, J.E. "External Economies and Diseconomies in a Competitive
Situation," Economic Journal, vol. 62 (March 1952), pp. 54-67.

Meiklejohn, D. "Liability for Oil Pollution Cleanup and the Water
Quality Improvement Act of 1970," Cornell Law Review, vol. 55
(July 1970), pp. 973-991.

Mendelsohn, A.I. "Maritime Liability for Oil Pollution — Domestic and
International Law," George Washington Law Review, vol. 38
(October 1969), pp. 1-31.

Mishan, E.J. "Reflections on Recent Developments in the Concept of
External Effects," Canadian Journal of Economics and Political
Science, vol. 31 (February 1965), pp. 3-34.

Mishan, E.J. "Welfare Criteria for External Effects," American Economic Review, vol. 51 (September 1961), pp. 594-613.

Modelski, G. "Communism and the Globalization of Politics," International Studies Quarterly, vol. 12 (December 1968), pp. 380-393.

Muskie, E.S. "Torts, Transportation, and Pollution: do the old shoes fit?," Harvard Journal on Legislation, vol. 7 (May 1970), pp. 477-494.

Nanda, V.P. "The Torrey Canyon Disaster: Some Legal Aspects," Denver Law Journal, vol. 44 (Summer 1967), pp. 400-425.

Nanda, V.P. and Stiles, K.R. "Offshore Oil Spills: An Evaluation of Recent United States Responses," San Diego Law Review, vol. 7 (July 1970), pp. 519-540.

Nelson-Smith, A. "The Effects of Oil Pollution and Emulsifier Cleansing on Shore Life in Southwest Britain," Journal of Applied Ecology, vol. 5 (April 1968), pp. 97-107.

Neuman, R.H. "Oil on Troubled Waters: The International Control of Marine Pollution," Journal of Maritime Law and Commerce, vol. 2 (January 1971), pp. 349-361.

Nixon, R.M. "Offshore Oil Production," Department of State Bulletin, vol. 62 (June 15, 1970), pp. 754-759.

O'Connell, D.M. "Continental Shelf Oil Disasters: Challenge to International Pollution Control," Cornell Law Review, vol. 55 (November 1969), pp. 113-128.

O'Connell, D.M. "Reflections on Brussels: IMCO and the 1969 Pollution Conventions," Cornell International Law Journal, vol. 3 (Spring 1970), pp. 161-188.

"Oil Pollution of the Sea," Harvard International Law Journal, vol. 10 (Spring 1969), pp. 316-359.

Pell, C. "The Oceans: Man's Last Great Resource," Saturday Review, vol. 52 (October 11, 1969), pp. 19-21, 62-63.

Parker, D.S. and Crutchfield, J.A. "Water Quality Management and the Time Profile of Benefits and Costs," Water Resources Research, vol. 4 (April 1968), pp. 233-246.

Prest, A.R. and Turvey, R. "Cost-Benefit Analysis: A Survey," Economic Journal, vol. 75 (December 1965), pp. 683-735.

"Proposal to Protect Maine from the Oilbergs of the 1970's," University of Maine Law Review, vol. 22 (1970), pp. 481-510.

Ranken, M.B.F. "Can We Delay the Next Major Tanker Disaster?," Ocean Industry, vol. 6 (June 1971), pp. 35-39.

Read, J.E. "The Trail Smelter Dispute," The Canadian Yearbook of International Law, 1963. Vancouver: Publications Centre, University of British Columbia and Canadian Branch, International Law Association (1963), pp. 213-229.

Reich, C.A. "The New Property," Yale Law Journal, vol. 73 (April 1964), pp. 733-787.

Rempe, G.A., III. "International Air Pollution — United States and Canada — A Joint Approach," Arizona Law Review, vol. 10 (Summer 1968), pp. 138-147.

Robinson, G.S. "The Regulatory Prohibition of International Supersonic Flights," International and Comparative Law Quarterly, vol. 18 (October 1969), pp. 833-846.

Ross, W.M. "The Management of International Common Property Resources," Geographical Review, vol. 61 (July 1971), pp. 325-338.

Schachter, O. and Serwer, D. "Maritime Pollution Problems and Remedies," American Journal of International Law, vol. 65 (January 1971), pp. 84-111.

Schaefer, M.B. "Some Recent Developments Concerning Fishing and the Conservation of the Living Resources of the High Seas," San Diego Law Review, vol. 7 (July 1970), pp. 371-407.

Scitovsky, T. "External Diseconomies in the Modern Economy," Western Economic Journal, vol. 4 (Summer 1966), pp. 197-202.

Scitovsky, T. "Two Concepts of External Economies," Journal of Political Economy, vol. 62 (April 1954), pp. 143-151.

Sewell, W.R.D. "Multiple-Purpose Development of Canada's Water Resources," Water Power, vol. 14 (April 1962), pp. 146-151.

Shutler, N.D. "Pollution of the Sea by Oil," Houston Law Review, vol. 7 (March 1970), pp. 415-441.

Singleton, J.F. "Pollution of the Marine Environment from Outer Continental Shelf Operations," South Carolina Law Review, vol. 22 (Spring 1970), pp. 228-240.

Snoke, K.P. "Admiralty Law: California Sues a Vessel in Rem for Oil Discharge Damages to its Water and Marine Life," Tulsa Law Journal, vol. 6 (August 1970), pp. 257-263.

Stein, M. "Aspects of United States and Canadian Water Law," University of Toronto Law Journal, vol. 20 (1970), pp. 69-71.

Sugden, D.E. "Piping Hot Wealth in a Sub-zero Land," Geographical Magazine, vol. 44 (January 1972), pp. 226-229.

Sweeney, J.C. "Oil Pollution of the Oceans," Fordham Law Review, vol. 37 (December 1968), pp. 155-208.

Thompson, A.R. "Basic Contrasts between Petroleum Land Policies of Canada and the United States," University of Colorado Law Review, vol. 36 (Winter 1964), pp. 187-221.

Tripp, J.T.B. and Hall, R.M. "Federal Enforcement Under the Refuse Act of 1899," Albany Law Review, vol. 35 (Fall 1970), pp. 60-80.

Tucker, A.J. "Boom in Tankers Ahead," Ocean Industry, vol. 5 (January 1970), pp. 35-39.

Ullman, E.L. "Political Geography in the Pacific Northwest," Scottish Geographical Journal, vol. 54 (July 1938), pp. 236-239.

Utton, A.E. "Protective Measures and the Torrey Canyon," Boston College Industrial and Commercial Law Review, vol. 9 (1968), pp. 613-632.

Weissberg, G. "International Law Meets the Short-Term National Interest: The Maltese Proposal on the Sea-Bed and Ocean Floor — Its Fate in Two Cities," International Comparative Law Quarterly, vol. 18 (January 1969), pp. 41-102.

Wildavsky, A. "Aesthetic Power or the Triumph of the Sensitive Minority over the Vulgar Mass, A Political Analysis of the New Economics," Daedalus, vol. 96 (Fall 1967), pp. 1115-1128.

Wilkes, D. "International Administrative Due Process and Control of

Pollution — The Canadian Arctic Waters Example," Journal of Maritime Law and Commerce, vol. 2 (April 1971), pp. 499-539.

Williams, B.R. "Economics in Unwonted Places," Economic Journal, vol. 75 (March 1965), pp. 20-30.

Wollman, N. "The New Economics of Resources," Daedalus, vol. 96 (Fall 1967), pp. 1099-1114.

Wolman, A. "Pollution as an International Issue," Foreign Affairs, vol. 47 (October 1968), pp. 164-175.

Wulf, N.A. "International Control of Marine Pollution," Judge Advocate General Journal, vol. 25 (December 1970-January 1971), pp. 93-100.

Reports and Proceedings

American Petroleum Institute. Prevention and Control of Oil Spills, Proceedings of a Joint Conference, June 15-17, Washington, D. C. Washington: American Petroleum Institute (1971).

Battelle Memorial Institute, Pacific Northwest Laboratories. Oil Spillage Study: Literature Search and Critical Evaluation for Selection of Promising Techniques to Control and Prevent Damage; for Department of Transportation, United States Coast Guard, Washington, D.C. Washington: United States Department of Commerce, Clearinghouse for Federal Scientific and Technical Information (1967).

Canadian Council of Resource Ministers. Background Papers Prepared for the National Conference on "Pollution and Our Environment," Montreal, October 31-November 4, 1966. Montreal: Canadian Council of Resource Ministers (1967), 3 vols.

Canadian Council of Resource Ministers. A Digest of Environmental Pollution Legislation in Canada. Report prepared for and under the responsibility of Canadian Industries Limited. Montreal: Canadian Council of Resource Ministers (1970), 2 vols.

Chevalier, M. Social Science and Water Management: A Planning Strategy. Paper prepared for the Policy and Planning Branch, Department of Energy, Mines, and Resources. Ottawa: Queen's Printer (1970).

Commission on Marine Science, Engineering, and Resources. Panel Reports. Washington: United States Government Printing Office (1969), 3 vols.

Crutchfield, J.A., Bish, R.L., Maleng, N.K., and Warren, R.O. "Socioeconomic, Institutional, and Legal Considerations in the Management of Puget Sound," Interim Report Submitted to the National Science Foundation, Sea Grant No. NE-RD-P11-1. Seattle: University of Washington (September 30, 1969), typewritten.

Davis, J. Canadian Energy Prospects. Special study for the Royal Commission on Canada's Economic Prospects. Ottawa: Queen's Printer (1957).

Hawkes, A.L. "A Review of the Nature and Extent of Damages Caused by Oil Pollution at Sea," Transactions North American Wildlife Conference, vol. 26 (1961), pp. 343-355.

Puget Sound Task Force, Comprehensive Water Resources Study. "Navigation: Puget Sound and Adjacent Waters," Preliminary Draft (1970), mimeo.

Resources for Tomorrow Conference. Proceedings of the Conference held in Montreal, 1961. Ottawa: Queen's Printer (1962), 3 vols.

Seattle Chamber of Commerce, Marine Exchange. Annual Reports. (yearly since 1949).

Sewell, W.R.D., Judy, R.W., and Ouellet, L. Water Management Research: Social Science Priorities. Papers prepared for the Policy and Planning Branch, Department of Energy, Mines, and Resources. Ottawa: Queen's Printer (1969).

Thorne, C.M. "How Can the People of the State of Washington Coexist with the Oil Industry?," report submitted to the Oceanographic Institute of Washington (December 1, 1970), typewritten.

Urbanization and Natural Environment, report of a conference held at the University of British Columbia, December 13 and 14, 1968. Vancouver and Seattle: University of British Columbia, School of Community and Regional Planning and the University of Washington, Department of Urban Planning (1969).

Water Pollution as a World Problem: the Legal, Scientific, and Political

Aspects, report of a conference held at the University College of Wales, Aberystwyth, July 11-12, 1970. London: Europa Publications for the David Davies Memorial Institute of International Studies (1971).

Books

Alexander, L.M. (ed.) The Law of the Sea. Columbus: Ohio State University Press (1967).

Berber, F.J. Rivers in International Law. New York: Oceana Publications (1959).

Bloomfield, L.M. and Fitzgearld, G.F. Boundary Water Problems of Canada and the United States. Toronto: Carswell Press (1958).

Chacko, C.J. The International Joint Commission between the United States and the Dominion of Canada. New York: Columbia University Press (1932).

Chamberlain, J.P. The Regime of the International Rivers, Danube and Rhine. New York: Columbia University Press (1923).

Chapman, J.D. (ed.) The International River Basin. Vancouver: Publications Centre, University of British Columbia (1963).

Christy, F.T., Jr. and Scott, A. The Common Wealth in Ocean Fisheries, Some Problems of Growth and Economic Allocation. Baltimore: Johns Hopkins Press (1964).

Cooley, R.A. and Wandesforde-Smith, G. (eds.) Congress and the Environment. Seattle and London: University of Washington Press (1970).

Cowan, E. Oil and Water: The Torrey Canyon Disaster. Philadelphia: Lippincott (1968).

Crutchfield, J.A. and Pontecorvo, G. The Pacific Salmon Fisheries: A Study in Irrational Conservation. Baltimore: Johns Hopkins Press (1969).

Dales, J.H. Pollution, Property and Prices. Toronto: University of Toronto Press (1968).

Darling, F.F. and Milton, J.P. (eds.) Future Environments of North America. Garden City, New York: The Natural History Press (1966).

Degler, S.E. (ed.) Oil Pollution: Problems and Policies. Washington: Bureau of National Affairs Book (1969).

Degler, S.E. and Bloom, S.C. Federal Pollution Control Programs: water, air, and solid wastes. Washington: Bureau of National Affairs Book (1969).

Dolman, C.E. (ed.) Water Resources of Canada. Toronto: Royal Society of Canada, Studia Varia, no. 11 (1967).

Dooley, P.C. Elementary Price Theory. New York: Appleton-Century-Crofts (1967).

Duesenberry, J.S. Income, Saving and Theory of Consumer Behavior. Cambridge: Harvard University Press (1949).

Firey, W. Man, Mind, and Land — A Theory of Resource Use. Glencoe: The Free Press of Glencoe, Illinois (1960).

Frankel, P.H. Essentials of Petroleum: a key to all economics. London: Cass (1969).

Galbraith, J.K. The Affluent Society. Toronto: Mentor Books (1963).

Gorove, S. Law and Politics of the Danube. The Hague: Martinus Nijhoff (1964).

Hartshorn, J. Oil Companies and Government. London: Faber and Faber (1962).

Hartshorn, J. Politics and World Oil Economics. New York: Frederick A. Praeger (1967).

Herfindahl, O.C. and Kneese, A.V. Quality of the Environment: An Economic Approach to Some Problems in Using Land, Water and Air. Baltimore: Johns Hopkins Press (1965).

Hoult, D.P. (ed.) Oil on the Sea. New York: Plenum Press (1969).

Jarrett, H. (ed.) Environmental Quality in a Growing Economy. Baltimore: Johns Hopkins Press (1966).

Jarrett, H. (ed.) Perspectives on Conservation. Baltimore: Johns Hopkins Press (1958).

Kapp, K.W. The Social Costs of Private Enterprise. New York: Schocken (1970).

Kneese, A.V. The Economics of Regional Water Quality Management. Baltimore: Johns Hopkins Press (1964).

Kneese, A.V. Water Pollution: Economic Aspects and Research Needs. Baltimore: Johns Hopkins Press (1962).

Kneese, A.V. and Bower, B.T. Managing Water Quality: Economics, Technology, Insititutions. Baltimore: Johns Hopkins Press (1968).

LaForest, G.V. Natural Resources and Public Property Under the Canadian Constitution. Toronto: University of Toronto Press (1969).

Landsberg, H.H., Fischman, L.L. and Fisher, J.L. Resources in America's Future. Baltimore: Johns Hopkins Press (1963).

Lederman, W.R. (ed.) The Courts and the Canadian Constitution. Toronto: McClelland and Stewart, Carleton Library Series, no. 16 (1964).

MacNeil, J.W. Environmental Management. Ottawa: Information Canada (1971).

McDougal, M.S. and Burke, W.T. The Public Order of the Oceans. New Haven: Yale University Press (1962).

Marx, W. The Frail Ocean. New York: Ballantine Books (1967).

Mishan, E.J. The Costs of Economic Growth. London: Staples (1967).

Moss, J.E. Character and Control of Sea Pollution by Oil. Washington: American Petroleum Institute (1963).

Rogers, G.W. (ed.) Change in Alaska — People, Petroleum, and Politics. College, Seattle and London: University of Alaska Press and University of Washington Press (1970).

Sherman, P. Bennett. Toronto: McClelland and Stewart (1966).

249

Smith, H.A. Federalism in North America. Boston: Chipman Law Publishing Company (1923).

Tugendhat, C. Oil: The Biggest Business. New York: G.P. Putnam (1968).

University of British Columbia, Faculty of Law. University of British Columbia Law Review, vol. 6 (June 1971).

University of California, Davis. University of California, Davis Law Review, vol. 1 (1969).

University of New Mexico, School of Law. Natural Resources Journal, vol. 11, April, 1971 (July 1971).

Vagner, J. (ed.) Oil on Puget Sound. Seattle: University of Washington Press (1972).

Ward, B. Spaceship Earth. New York: Columbia University Press (1966).

Whyte, W.H. The Last Landscape. New York: Doubleday (1968).

International Publications

Heeney, A.D.P. and Merchant, L.T. Canada and the United States: Principles for Partnership. Ottawa: Queen's Printer (1965).

Intergovernmental Maritime Consultative Organization. Preliminary Report to an International Subcommittee (Torrey Canyon) of the International Maritime Committee. London: IMCO (1967).

International Conference on Oil Pollution of the Sea. Proceedings. Conference held in Rome, October 7-9, 1968. Distributed by Warren and Son Ltd., Winchester, United Kingdom (1968).

International Joint Commission. Rules of Procedure and Text of Treaty. Ottawa, Canada and Washington, D.C.: International Joint Commission and United States Government Printing Office (1965).

International Joint Commission. Special Report on Potential Oil Pollution, Eutrophication, and Pollution from Watercraft. Ottawa and Washington, D.C.: The International Joint Commission (1970).

Organization for Economic Co-operation and Development, Special

Committee for Oil. Pipelines in the United States and Europe and Their Legal and Regulatory Aspects. Paris: OECD (1965).

United Nations Treaty Series (1960-1970).

Newspapers and Periodicals

Bureau of National Affairs, Environmental Reporter, 1970-1971.

Conservation Foundation, Conservation Newsletter, 1969-1972.

Financial Post, 1969-1972.

International Legal Materials, vols. 1-10, 1962-1971.

National Wildlife Federation, Conservation Report, 1969-1972.

New York Times, 1966-1972.

Saturday Review, August 7, 1971.

Seattle Post-Intelligencer, 1969-1972.

Seattle Times, 1969-1972.

Vancouver Province, 1969-1972.

Vancouver Sun, 1969-1972.

Personal Communications

Interviews

Buchanan, H.O., Director, Marine Services, Western Region, Canada, Department of Transport, Vancouver. May 17, 1971.

Martin, Sidney S., Marine Manager, Imperial Oil Limited, Vancouver. January 28, 1971.

Phaups, James, Lieutenant, Oil Pollution Officer, Thirteenth Coast Guard District, Seattle. April 15, 1971.

Weymouth, Edward, Executive Secretary, Western Oil and Gas Association, Seattle. January 19, 1971.

Letters

Brower, Charles N., Assistant Legal Advisor, Department of State, Washington, D.C. June 17, 1970.

Cooch, F.G., Staff Specialist, Migratory Bird Populations, Ottawa. May 12, 1971.

MacCallum, J.L., Q.C., Assistant to the Chairman and Legal Advisor, Canadian Section, International Joint Commission. May 8, 1970.

Macgillivray, R.R., Director, Marine Regulations, Department of Transport, Ottawa. December 17, 1971.

O'Riordan, Timothy, Department of Geography, Simon Fraser University. June 1, 1970.

Sharp, Mitchell, Secretary of State for External Affairs, Ottawa. March 2, 1971.

Williston, Ray, Minister of Lands, Forests, and Water Resources, Victoria, British Columbia. April 16, 1970.

Willman, James C., Chief, Oil Pollution and Hazardous Materials, Environmental Protection Agency, Northwest Region, Portland, Oregon. March 10, 1971.

ADDENDUM

Christy, F.T., Jr. "Economic Criteria for Rules Governing Exploitation of Deep Sea Minerals," International Lawyer, vol. 2 (January 1968), pp. 224-242.

Gordon, H.S. "The Economic Theory of a Common Property Resource; The Fishery," Journal of Political Economy, vol. 62 (April 1954), pp. 124-142.

INDEX

A

Absolute liability, 77, 169, 186; under Civil Liability Convention (1969), 166. See also Liability and Strict liability.

Acts of God: 89, 186

Admiralty Court, 123

Admiralty law, 105

Africa, 1

Agreement Concerning Pollution of the North Sea by Oil, 172

Air pollution, 2, 177; international problems of, 25; in Windsor-Detroit area, 195

Alaska: discovery of oil in, 26; oil deliveries from, 130

Alaska oil, 69-71, 193, 205, 214, 222; problem of, 69-71; volume of, 70-71; and Canadian claims, 168; litigation concerning, 212

Alaska Pipeline, 173, 217-219, 223; and Cherry Point spill, 222

Alberta, 62

Anacortes, 203; oil spill near, 67, 212, 220

Anderson, David, 218-219; suit of, 212n

Anti-pollution agreements, 5

Anti-pollution laws. See Laws, Liability, and under name of specific law.

Arbitration: under Public Law Convention, 164; and Canada, 165

Arctic, 70; threat to, by voyage of Manhattan, 19; Canadian pollution control zone for, 23; pollution of, 24

Arctic Waters Pollution Prevention Act, 101, 120-121, 207, 224

Aromatic content, 8-9

Arrow, 101, 127, 204; costs of, to Canadian government, 10; sinking of, 19

Aspin, Representative Les, 218

Atlantic Richfield, 71, 220, 222; tankers constructed by, 216; and Cherry Point Spill, 224-226

Audubon Society, 66

Auks, 8

B

Ballast, 216

Baltic Sea, 196

Barclay Sound, 214

Barges: and spillage, 1; in Puget Sound and Strait of Georgia, 65

Bator, F.M., 36, 37

Battelle Memorial Institute, 17

Bay of Fundy, 195

Bilateral agreements, 5

Bilateral organization: to control pollution, 189; ways of accomplishment, 191; composition of, 191; research branch of, 191; financing of, 191; functions of, 191-192; and shipping restrictions, 191-192

Bilge pumping, 7, 67

Bellingham, Washington, 67, 203

Birch Bay, 222

Birch Point, 222

Corfu Channel Case, 77

Costs, 50, 51; transfer of, 2; aesthetic costs, 10; opportunity costs, 35; cost-benefit analysis, 36; of pollution, 40; social, 40; downstream, 40; failure to account for, 41; of waste disposal, 51; of petroleum products, 187. See also Marginal costs.

Council on Environmental Quality, 227

Courts: of British Columbia, 77; ICJ, 77; of Washington, 77; Supreme Court, 218

"Cowboy" economy, 4

Crescent Beach, 222

Crowe, B., 6

Crude oil, 10, 62, 64; as pollution source, 7; from Alaska, 69; transport of, 70, 220; prices of, 205; production of, 220

Crutchfield, J.A., 51-52, 138

Cunningham, P., 94

D

Dales, J.H., 30-31

Damages, 2, 175; under Civil Liability Convention (1969), 166; absolute liability for, 186; strict liability for, 186; litigation of, 187. See also Liability, Absolute liability, and Strict liability.

Davis, Jack, (Minister of the Environment), 15, 225, 227

Department of the Interior. See U.S. Department of the Interior.

Detection: of oil spills, 6

Detroit, Mich., 177, 195

Diesel fuel, 7

Domestic laws, 179; to control pollution, 158; in U.S. and Canada, 185

Downstream costs, 40

Drainage basins, 210

Drilling, 16, 135, 137; increases in, 1; spills resulting from, 17; off Pacific Coast, 26; and marine environment, 65; in Puget Sound, 65–66; in Strait of Georgia, 67, 104; on continental shelf, 161–162; hazards, 185. See also Offshore drilling.

Drydock, 221

Dungeness Light, 216

E

Ecology, 4, 5, 33–34, 207; and standards for waste discharge, 50

Economic costs, 41–42

Economic feasibility, 34

Economic rent, 41

Economic values, 33

Ecosystems, 8

Energy, 10, 12–13

Enforcement: of anti-pollution laws, 6, 220–221; of Geneva Convention (1958), 162

English Channel, 196

Environment, 19, 22, 35, 41

Environmental Protection Agency, 220

Esso Petroleum Company Ltd. vs. Southport Corporation, 106

Europe: and spillage from tankers, 1; increased oil demand, 16

Evans, Governor, 226

Exploitation, 41

Exploration: increases in, 1, 16; in Puget Sound, 65–66

Extension of Admiralty Jurisdiction Act of 1948, 80

Externalities, 4, 35, 38–41, 47–48, 54n, 55n, 172, 185, 228

Extraction, 41

Extraterritorial oil damage, 133–134, 136; Canadian laws concerning, 108; in Puget Sound and Strait of Georgia, 130; and Civil Liability Convention (1969), 167

Extraterritorial oil pollution, 101, 104–105, 139

F

Federal Water Pollution Control Act of 1948, 81, 84, 87

Federal Water Pollution Control Administration, 70

Financial responsibility, 88–89, 166–167

Financing: methods of, 187; of pollution control, 186, 194

Firey, Walter, 33, 205

Fischman, Leonard L., 13

Fish: affected by oil spills, 8, 11; and detergents, 9; fishing intensity, 42–43; damages paid for, 94

Fisher, Joseph L., 13

Fisheries Act, 108, 109

Fishing, 41–42, 61; commercial fisheries, 61–62; and economic efficiency, 43; and pollution damage claims, 123–124

Flag states, 85, 159, 160

Florida, 91

Food chain, 8

Foreign nationals, 219

France, 10, 157

Fraser River system, 58, 190

Freedom of the seas: favored, 156; in multilateral agreements, 207; and Canadian legislation, 224

Free enterprise marked: and environmental quality, 35; as allocator of resources, 35. See also Market economy.

Friends of the Earth, 218

G

Gasolene, 7

Gastineau, 214

Geneva Convention on the Continental Shelf of 1958, 161-162

Geneva Convention on the High Seas, 162

Geneva Conventions: signers of, 163; violations of, 163. See also under names of individual conventions.

Germany, Federal Republic of, 157

Gordon, H., 41, 42-47

Great Britain, 10

Great Lakes, 177; pollution control report on, 157; studies by International Joint Commission, 175; pollution of, 195; and Regional Commission, 193

Greece: shipping practices challenged, 224

Gulf Oil Company, 66

Gulls, 8

H

Harbor Advisory Radar System, 100

Hardin, G., 4

Haro Strait, 71

Hart, Judge George, 218

Hatshorn, J.E., 10

Health Act, 108, 116

Heating oil, 7

Heeney, Arnold, 174

High seas, 23; pollution of, 162; regulations concerning, 162

Honeywell, 69; oil spill by, 219

Hovanesian, A., 135-136, 188; discusses regional organization, 188

House of Commons, 223

Hovercraft, 221

Howe Sound, 67

Hydrocarbons, degradation of, 8

Hydroelectricity, 15

I

Idaho, 65

IMCO. See Intergovernmental Maritime Consultative Organization

Industrial oil, 7

Industries, 4

International organizations, 6, 155, 188

International pollution, 5; Canadian response to, 101

International treaties: between Canada and U.S., 185. See also under names of individual treaties.

International tribunals, 76-78

International water quality management, 34

International Water Quality Management Area, 225

J

Jackson, Senator Henry, 226

Jackson, R.I., 157-158

Japan, 70, 224

Jay, J., 5

Johnstone Straits, 51; and petroleum trade, 65

Joint Commission. See Bilateral Commission.

Jones Act, 217; and oil transportation, 70

Jurisdiction, 208-209; of state and federal government, 79

K

Kerosene, 7

Kneese, A.V., 40, 48, 50, 51, 138

L

Landsberg, Hans H., 12

Lake Erie, 176

Lake Lanoux case, 77

Lake Ontario, 177

Law of the Sea Conference, 224, 228

Laws, 23-25, 77-78, 80-100; and oil pollution, 72-76; common law
    principles, 80; enforcement of, 220-221

Legal action: right of, 185; in U.S. courts, 219

Legault, L.H.J., 159

Legislative process: Canadian, 209; U.S., 209; reform of, 210

Legislative proposals, 224

Liability, 79-80, 88, 89, 91, 93, 186; under Water Quality Improvement
    Act of 1970, 89; state liability, 76; liability standards, 79-80;
    and financial responsibility, 88-89; third party liability, 89; and
    Civil Liability Convention (1969), 165-168; limitations on, 166-
    167, 169, 186; limitations under TOVALOP, 171. See also
    Limited liability, Strict liability, and Absolute liability.

Liberia, 224

Limited liability, 166-167, 169, 186; legislation on, 89; disadvantages
    of, 187

Limited Liability Act of 1851, 89

Louisiana, 1, 34, 46

Luxembourg, 157

M

Magnusson, Senator Warren, 216, 226

Maine: laws against spills, 23, 25; liability re discharge, 91

Main-New Brunswick border, 195

Migratory birds: Migratory Birds Convention, 174; Migratory Birds Convention Act, 108, 110

Mullusca, 9

Monitoring devices, 6

Montreal, Canada, 195

Morton, Secretary Rogers, 218-219, 223

Multilateral treaties, 5, 207

N

Nanaimo, British Columbia, 214

Nanda, V.P., 192

National Contingency Plan (U.S.), 96-97

National Contingency Plan (Canada), 101

National Energy Board, 15

National Environmental Policy Act of 1969, 212n; impact of, 219

National security, 219

Natural gas: percentage use of, 10; properties of, 12; as precent of energy consumed, 13

Natural hydrocarbons, 9

Natural resources: allocational criteria, 205

Natural seepage, 6-7

Navigable Waters Protection Act, 108, 111

Navigable Waters Safety and Environmental Quality Act of 1972, 215-216

Negligence, 83, 101, 105, 106, 136, 137; in establishing liability, 80

Outer Continental Shelf Lands Act, 89

Ownership: of common property resources, 41; of the seas, 208

P

Pacific Northwest: as area of potential spillage, 26; boundary problems between U.S. and Canada, 72, 76

Pacific Halibut Commission, 174

Pacific Salmon Commission, 174, 190-191

Paish, Howard, and Associates, 219

Paish Report, 221

Parker, D.S., 51-52, 138

Persian Gulf, 172, 196

Persistant oil, 7

Petroleum: consumption of, 12; production of, 12, 15, 16; demand for, 13; Canada's energy needs for, 15

Petroleum and Natural Gas Act, 215

Point Whitehorn, 222

Pipelines, 16, 62-65; rupture of, 7; construction of, 62, 64; map of, 63

Plankton, 9

Polluters, identification of, 40

Pollution: of oceans, 1, 4, 5; occurence of, 2; of rivers, 2; of air, 2; international, 2, 3, 5, 76-77; laws concerning, 5, 132-133, 227; control of, 6, 25-26, 157, 185-189, 194, 203-204, 207-208, 210-211, 215; sources of, 7-8; effects on plants and wildlife, 8; costs of, 9-10, 40, 50; political problems emanating from, 47; in Puget Sound and Strait of Georgia, 60; limited liability for, 185; of

Great Lakes, 195; of drainage basins, 210. See also Air pollution, Water pollution, River pollution, and Oil pollution, etc.

Pollution Control Act of 1967, 108, 117-119

Port Angeles, 193

Portland, Me., 195

Portland, Oregon, 65

Ports and Waterways Safety Bill of 1971, 100, 215-216

President (U.S.), 87-88

Prosecution, 79, 159; under Refuse Act, 82; and Oil Pollution Act, 82; and flag states, 85

Provincial Disaster Fund, 227

Prudhoe Bay, 212n

Public intervention, 40

Public Law Convention (1969), 163-165. See also International Convention Relating to Intervention on the High Seas in Cases of Oil Pollution Casualties.

Public opinion, 203-204

Public welfare, 36

Puget Sound, 39, 58, 99, 131, 173-174, 179, 187, 193, 197-198, 204, and passim throughout the text; and oil spillage, 26, 37, 61-62, 67, 130, 139, 219; drilling in, 34, 65-66; map of, 59; estuarine dynamics of, 60; wildlife species in, 62; traffic in, 68-71, 100, 134, 192; jurisdiction of, 72; and Arctic Water Pollution Prevention Act, 121; control of pollution in, 189

R

Rainy River, 177

Red River, 177

Reference Re: Offshore Mineral Rights of British Columbia, 103-104

Refined products: as pollution source, 7; in Puget Sound, 69

Refineries, 96; on Puget Sound and Strait of Georgia, 62-64; Cherry Point, 71; location restrictions upon, 192-193; wind shelter of, 220

Refuse Act of 1899, 81-82, 87, 111

Regional programs, 211; Regional Response Center, 97; Regional Contingency plan, 97-98; Canadian Regional Task Force, 101; regional organizations, 157-158; regional agreements, 172-173; Regional Commission, 192-194, 196-197

Registered judgement, 223

Report of the Panel on Marine Resources, 12

Resources: allocation of, 35, 39-40, 51; priorities for allocating, 33-34; misallocation of, 40, 46, 48, 50, 136; management of, 207; extraction of, 41. See also Common property resources.

Rhine River, 157

Rivers and Harbors Act of 1899. See Refuse Act of 1899.

River pollution, 2, 156; international efforts to control, 2; and international law, 158. See also Pollution.

Roberts Bank, 26

Rosario Strait, 71

Ross Dam, 54n

S

Saint Croix River, 177

St. Lawrence River, 175

Salmon: and natural hydrocarbons, 9; runs entering Puget Sound, 61-62; traffic control during runs of, 192

Salvage stations, 221

San Francisco, California, 1

San Juan Islands, 39

Santa Barbara, California, 19

Schachter, O., 202

Sea ducks, 8

Seagate, 8

Seattle, Washington, 54n, 65

Seismic exploration, 66

Sewage, 60

Shell Oil Company, 66

Shipowners, 24; insurance plan, 171; favorable bias toward, 156

Shipping rules, 159

Ships: size of, 17; foundering of, 17; guidance systems for, 23; traffic, 68; collisions of, 69; and Jones Act, 70; construction requirements, 84-85, 86. See also Tankers.

Sierra Club, 222; and offshore drilling, 66

Skagit River: river system, 58; Skagit Valley-Ross Dam issue, 54n, 196

Skimmers, 216

Small operators, 187-188

Snohomish River, 58

Social costs, 48

Society for Pollution and Environmental Control (SPEC), 222

South America, 1

Sovereignty, 5, 76, 177, 205; and continental shelf, 161; and international resource management, 205; ownership of, 208; surrender of, 5, 6, 211

"Spaceman" economy, 4

Special regional fund, 187

Spills. See Oil spills

Standard Oil Refinery, 214

Stockholm Conference on the Human Environment, 3, 224-225, 226

Storage facilities, 7

Strait of Georgia, 39, 131, 173-174, 179, 187, 196-198, 204; and passim throughout text; spills in, 26, 37, 61-62, 67, drilling in, 61-62, 67; jurisdiction in, 58, 72, 104; map of, 59; estuarine dynamics of, 60; wildlife species in, 62; traffic in, 68, 100, 134, 192; and Arctic Water Pollution Prevention Act, 121; extraterritorial pollution in, 130, 139; and Cherry Point Spill, 222-224

Strait of Juan de Fuca, 58, 61-62, 71, 131

Strict liability, 83, 166, 186-187

Supreme Court, 218

Sunken vessels, 7

Superports, 17, 26

Svart, L.M., 203

Sweden, 172

Switzerland, 157

T

Tacoma, Washington, 65, 216

Tanker Owners Voluntary Agreement Concerning Liability for Oil Pollution (TOVALOP), 155, 173, 179, 207; reimbursement by, 171;
proof of negligence, 171; liability limits of, 171; increased
coverage of, 171-172; liability limits of, 172; modifications of,
207

Tankers, 16, 17; European spillage, 1; foundering of, 17; increased
size of, 17; "T-2" tankers, 17; capacities of, 17; in Strait
of Juan de Fuca, 61; in Puget Sound and Strait of Georgia, 65,
70; traffic level of, 68; ability of Puget Sound to handle, 71;
cleaning operations of, 84; ownership of, 156; restrictions upon
discharge from, 170; navigation control of, 192; double hull
requirements, 216; restrictions upon, 216; construction of, 216;
to supply Cherry Point refinery, 217

Tax: for pollution control, 186; to improve tankers, 188

Technology: role of, 6; to reduce spillage problems, 22; in resource
allocations, 34; and market economy, 36; and external diseconomies,
78, 138

Territorial Sea and Contiguous Zone Convention, 163

Texas Instruments, Inc., 212n

Texaco Oil Company, 66

Tongue Point, 222

Torrey Canyon, 9, 17, 18, 83, 156, 204; costs of, 10

TOVALOP. See Tanker Owners Voluntary Agreement Concerning Liability
for Oil Pollution.

Toxicosis, 8

Traffic control, 192

Trail Smelter Case, 76-77, 132, 136; and pollution damage, 76-77; and
jurisdiction, 76-77; as legal precednet, 77

Train, Russell, 227

Transboundary pollution, 185

Trans-Alaska Pipeline. See Alaska Pipeline.

Trans-Mountain Oil Pipe Line Company, 62, 64

Trespass, 101, 106; in establishing liability, 80

Trudeau, Prime Minister, P.E., 120

U

United Kingdom, 46

United Nations: efforts of, 3; and IMCO, 160

United States, 209; oil discharges and liability for spilled oil, 78; and
Public Law Convention (1969), 165; and Civil Liability Convention
(1969), 166; agreements with Canada, 174; responsiveness to
pollution problem, 203; and liability for pollution, 207; jurisdictional problems in, 208; courts of, 219; shipping practices, 224

U.S. Navy, 220

U.S. Coast Guard, 82, 83, 86, 92, 97, 98, 99, 100, 216, 220

U.S. Department of the Interior, 212n

U.S. President, 8; and action re oil spills, 88, 97

U.S. Army Corps of Engineers, 19

Used oil, 7

V

Vagners, Juris, 219-220

Valdez, 70, 212n

Vancouver Harbor, 214

Vancouver Island, 65, 68; transportation of oil to, 205; oil spill near,
214-215

Vancouver Province, 226

Vanlene, 214-215

Venezuela, 16

Victoria, British Columbia, 67

W

Wages, 35

Ward, Barbara, 3, 4, 54n

Washington, 26, 52, 60, 65, 68, 69, 76-77, 78, 204; and legislation re oil spillage, 23, 26, 27; industrial oil use in, 65; Department of Ecology of, 78-79, 92, 94, 95, 99, 220; pollution legislation, 86, 90-92, 95-96; liability laws, 91; concern with oil spills, 202; economy of, 203; prosecution re Cherry Point spill, 223

Washington Environmental Council, 66

Washington Oil Spill Act (1970), 92, 93, 94, 108, 137, 207

Water discharge: standards for, 49; and charge system, 50

Water circulation: in Puget Sound and Strait of Georgia, 61-62, 72

Water pollution: international laws on, 76, 158-159; Canada Laws concerning, 116, 117

Water Pollution Control Administration, 70, 92

Water Pollution Control Act, 1948, 84, 87

Water quality: social costs of, 51; management of, 57n; standards, 84; testing for, 157; and multilateral commissions, 157

Water Quality Improvement Act of 1970, 22-23, 81, 84, 87, 89-91, 93, 96-97, 99, 124-125, 132, 137; exemptions from, 87; liability provisions of, 89; assessment of, 89-91

Water resources: and market economy, 39; management of, 40

THE AUTHOR

Dr. William Michael Ross is a Canadian born in 1945 in Vancouver, British Columbia. He received his Bachelor of Education at the University of British Columbia in 1967 and Master of Arts Degree at the University of Toronto in 1968. Having completed his Doctoral degree at the University of Washington, Dr. Ross accepted a position as an assistant professor at Kent State University, Ohio.